Number 18 · THE CLIMATIC PHYSIOLOGY OF THE PIG

Volumes marked * are now out of print.

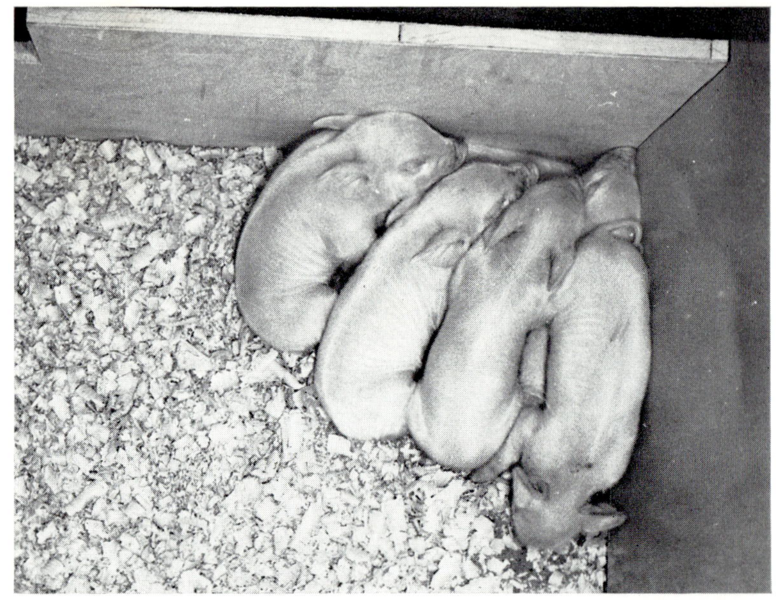

The response of a group of new-born pigs to a cold environment. In the upper photograph, at 15°C, the pigs are huddled together closely; in the lower photograph, in thermal neutrality at 35°C, the animals are spread out in relaxed postures.

THE CLIMATIC PHYSIOLOGY OF THE PIG

by

L. E. MOUNT

M.D., C.M.(McGill), M.B., B.S.(Lond.)

Head, Environmental Physiology Laboratory, Agricultural Research Council
Institute of Animal Physiology, Babraham, Cambridge

LONDON
EDWARD ARNOLD (PUBLISHERS) LTD

First published 1968

SBN: 7131 4140 9

Printed in Great Britain by
The Camelot Press Ltd., London and Southampton

CONTENTS

INTRODUCTION

I T is only during the last ten years or so that enough work has been done on the pig to allow a monograph of this sort to be written. Environmental physiology is in any event rather an all-embracing topic; in relation to the pig the subject tends to become even wider owing to the importance of the animal in agriculture and meat production and the increasing use of the pig in medical research.

Against this background, the present book is centred on the pig, and, where necessary, draws for species comparisons largely on sheep and man, which are of the same order of size and which provide the necessary links with the two main fields of applied interest referred to above. The work is not intended to be comprehensive, or to provide exhaustive discussions; it is written primarily from the standpoint of interest in thermal energy exchanges between the organism and its environment. Animals' thermal energy is, of course, provided by food, but nutrition is not considered except where necessary for the effects of environment to be explained. The emphasis is on experimental analysis of the pig's responses to its physical environment.

The book falls into four parts. The first of these comprises three chapters on environment, pigs, and energy exchange; there follow two chapters on thermal relations between the environment and the new-born and the older pig, constituting the second part. The third section of the book consists of four chapters on channels of heat exchange and the partition of heat loss. The fourth and final part includes commentaries on metabolic reference bases, behaviour, and pig farming, in three successive chapters.

In general, the practice of the *Journal of Physiology* has been followed in respect of symbols and references. The bibliography is necessarily selective, but those who may require a comprehensive coverage of publications concerned with the pig can obtain it in the 'Index of current research on pigs', an annual publication prepared by Dr R. Braude of the National Institute for Research in Dairying, Shinfield, Reading, England.

The pig has played a prominent part in man's affairs for a long time, probably about 7,000 years. On account of its rapid growth and omnivorous habit the animal has been an excellent source of dietary protein. It has also had considerable social importance in religion, symbolism, and heraldry; as an animal exerting either strong fascination or repugnance, it has given rise to expressions of praise or abhorrence (see Pope, 1962, for a general discussion of the pig's more abstract qualities and of the animal's relation to man). Few would deny the charm of young piglets, although the appeal of the large sow or boar is another matter. On the other hand, pigs have often been regarded as an incarnation of the devil, and are frequently used in the Bestiaries to symbolize a number of vices (Sillar & Meyler, 1961).

Terminology. The term 'pig' is commonly used in Britain, although in parts of America 'swine' is the general term. According to the *Shorter Oxford English Dictionary*, 'swine' is 'an animal of the genus *Sus* or family *Suidae* . . . now only literary or dialectal, or generic term in zoology'. 'Pig' is given as 'the young of swine: by extension, a swine of any age; any of various species of Suidae'. The two terms thus seem to have the same meaning, and the term 'pig' is used in this book. Where the term 'sow' is used, it implies a mature female pig which has had at least one litter; 'gilt' is the term applied to a nulliparous female. 'Boar' describes the intact male; the general practice with male pigs which are to be raised for meat production is to castrate them early in life, although there is some opinion in opposition to this. 'Farrowing' is the word used to describe parturition.

I am indebted to a number of people who have helped me greatly in the preparation of this book. It is not possible to name every individual, but I am particularly grateful to Miss C. O. Hebb for encouraging this work and for reading and criticizing some of the chapters; I am indebted for similar generous help to Drs B. A. Baldwin, J. Bligh, D. L. Ingram, and D. McK. Kerslake. Any errors or omissions in the book, however, remain my responsibility. Colleagues in the Environmental Physiology Laboratory have helped me in many ways; the tasks of obtaining many papers and other publications, and the preparation of figures, have been made simple for me by the unfailing efforts of the members of the Library Staff and Photographic Section at Babraham. I should like to express my thanks to Mrs V. Coulson and Miss C. Page for secretarial assistance.

Finally, I am grateful to the authors and publishers who have given me permission to reproduce material from their work.

L. E. MOUNT

Babraham, Cambridge
October, 1967

I

CLIMATE AND ENVIRONMENT

CLIMATE is weather generalized in space and time. A given climate applies to a given region of the earth, so that there are climates which are hot or cold, wet or dry; different climates characterize the overall environments in which plants and animals live. The term 'climate' is used in this sense to refer to the *macro-climate*, which may be defined as those conditions obtaining in the atmosphere at large, including annual rainfall, number of days in the year with wind-speed exceeding a given level, mean number of hours of sunshine, cloudiness, and so on.

The macro-climate of a region influences to a large degree the *meso-climate* of a smaller zone, for example the southern as opposed to the northern slope of a hill, or the lee side of a wooded

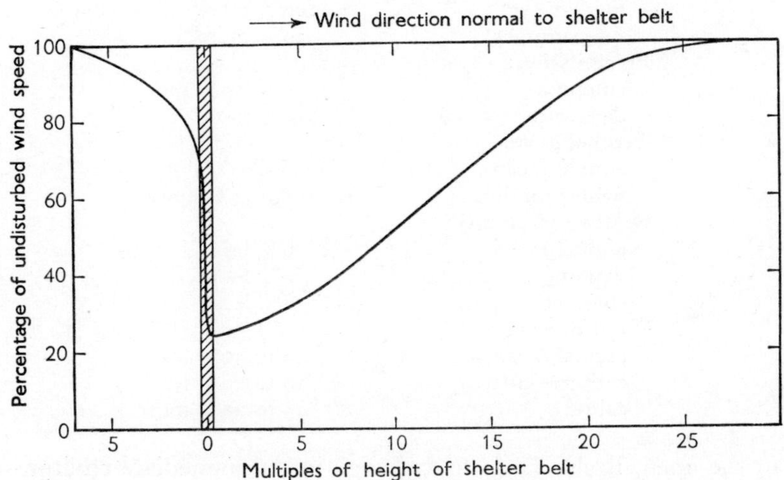

FIG. 1.1. The reduction of wind-speed by a shelter belt. Ordinate: percentage of undisturbed wind-speed; abscissa: distance in multiples of height of the shelter belt (from Landsberg, 1964, by permission of Elizabeth Licht).

belt as opposed to an exposed plain. Such a modification of macro-climate is produced by the common use of a belt of trees as a wind-break, with diminished wind-speeds extending for considerable distances down-wind (Fig. 1.1). Not only does this affect animals

TABLE 1.1

Climatic changes produced by cities (from Landsberg, 1964, by permission of Elizabeth Licht)

Element	Compared with rural environs
Contaminants:	
dust particles	10 times more
sulphur dioxide	5 times more
carbon dioxide	10 times more
carbon monoxide	25 times more
Radiation:	
total on horizontal surface	15 to 20% less
ultraviolet, winter	30% less
ultraviolet, summer	5% less
Illumination:	
visible light, summer	5% less
visible light, winter	15% less
Cloudiness:	
clouds	5 to 10% more
fog, winter	100% more
fog, summer	30% more
Precipitation:	
amounts	5 to 10% more
days with $\leqq 5$ mm	10% more
Temperature:	
annual mean	0·5 to 1°C more
winter minima	1 to 2°C more
Relative Humidity:	
annual mean	6% less
winter	2% less
summer	8% less
Wind Speed:	
annual mean	20 to 30% less
extreme gusts	10 to 20% less
calms	5 to 20% more

in the open, it also leads to a change in the immediate environ-ment of a house or shelter, and thus indirectly affects the occupants. Stoekeler & Williams (quoted by Landsberg, 1964) estimated at 20–30% the reduction of fuel consumption due to wind-breaks round farm houses on the American plains. The collection of buildings in towns and cities modifies the environment in various ways; some of the effects produced by cities are indicated in Table 1. The corresponding effects of the town environment in decreasing the heating requirements of houses are illustrated in Fig. 1.2.

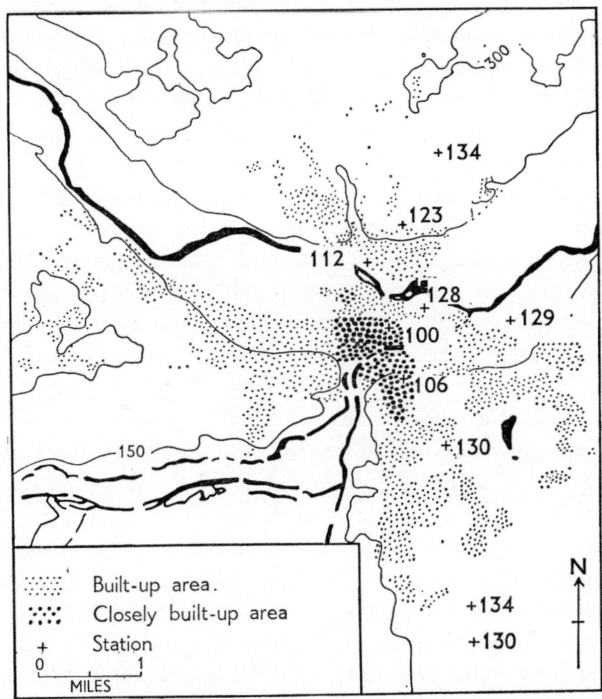

FIG. 1.2. Proportional heat losses from average houses in the Reading (England) area for the twelve-month period December 1951 to November 1952. The differences due to elevation are only 5 or 6%; the 150-foot contour line is marked (from Parry, 1957, by permission of *Advancement of Science*).

THE MICRO-ENVIRONMENT

The *micro-climate* comprises the conditions immediately around the given animal or object under consideration, and in this way partially defines the animal's *micro-environment* as a compound attribute of those environmental quantities which impinge on the animal, and which fluctuate both with time and with the animal's behaviour. The reaction of the animal to the environment is generally such as to result in an effective micro-climate most in keeping with its 'comfort', that is a situation in which external demands on the animal are reduced to a minimum. Under hot conditions an animal will lie down, relaxed, on cool ground if this is available; in the cold, the same animal will either stand, or rest on the ground with a minimum of contact, thus behaving in a manner which leads to a reduction in conductive heat loss. Moving behind a wind-break, and seeking shade from solar radiation, will reduce convective heat loss and the radiant heat load respectively; keeping dry reduces evaporative heat loss to that due to the animal's endogenous mechanisms of sweating and respiratory heat loss.

It would seem that the most appropriate way in which the thermal micro-environment can be defined is therefore in terms of an animal's heat exchanges through the several channels of conduction, convection, radiation, and evaporation (see Chapter 9), and in terms of the factors influencing these exchanges. Thus conductive heat exchange is influenced by posture and immediate contact with the surroundings; radiant heat exchange is also influenced by posture, but it may take place with heat sources which may be at a great distance, as with the sky and sun. Convective and evaporative heat exchanges are determined by local variations in air temperature, wind speed, and humidity. The combination of behavioural responses and apparently minor physical features of the environment can affect very considerably the physiological responses normally necessary for the animal's thermal stability or even survival; even out on the hillside sheep find those areas where the immediate wind-speed is lower, and where conditions are warmer (Blaxter, 1964). In this connection, Cresswell and Thomson (1964) suggested from measurements of wind-speed at different heights above the ground that even so apparently minor a feature as a tussocky pasture affords shelter to sheep, not only from wind but possibly also from blown rain and snow.

Thus the micro-environment is the immediately effective environment of the organism, and measurements with meteorological instruments which are placed at sites remote from the animal studied could produce large errors in computing its heat exchanges. There have been attempts to relate macro- and micro-climate in given situations, suc has valleys and forests. The term 'micro-climate' used in this sense approaches the meaning of 'meso-climate', and is taken to mean the layer of air which is altered to a considerable degree by the surface of the earth or other surfaces, as for example by trees (Holmes & Dingle, 1965). The scale of size of micro-climates extends from millimetres to metres, and if it were possible to obtain useful quantitative relations between macro- and micro-climates it would then be possible to predict highly localized climatic situations from the general weather information gained by weather stations, with obvious advantages to agriculture. An animal, however, can produce much greater variation than a plant in its exposure to features of its micro-environment by virtue of postural or other behavioural changes, so that it would be difficult to infer an animal's micro-climate solely from the macro-climate and the local physical terrain, as might be done with vegetation.

The micro-climate of plants is the result of exposure to external heat sources and sinks; micro-climates of mammals and birds are affected to a large degree by their own metabolic heat production, as a result of which they can regulate their body temperatures. The effect of vegetation on its own micro-climate is more limited and like the effect of an animal's hair coat—it influences insulation and the exchange of heat, but does not produce large quantities of heat. The reactions of the animal, on the other hand, exert the predominant influence on its effective micro-climate. A very clear example of this is provided by the considerable modifying influence on a cold environment of pigs huddling together; the effectiveness of this is seen in the considerable decrease in the energy expenditure which they require in order to maintain a normal body temperature (see Chapter 5). In contrast to plants, therefore, animals have considerable control over their micro-environment.

CRYPTO-CLIMATE

The climate inside a house, tent, or cave is sometimes termed the *crypto-climate* (Landsberg, 1964), a special case of the *eco-climate* which applies to a given area. The establishment of suitable crypto-climates is now of increasing importance not only for people but also for the growing numbers of farm animals which are housed under conditions of intensive husbandry; clearly it is necessary to determine the environment which is most suitable for the health, welfare, and productivity of various classes of livestock. The immediate environment of housed animals, however, is not simply that inside the house; again it is the modified micro-environment, which is influenced to a large degree by the animals themselves. The location, orientation, and structure of the house have considerable influence on heat exchange between the house and the general surroundings, and consequently on the internal crypto-climate which initially determines the animal's responses to its environment. The combination of these responses and the crypto-climate then produces the micro-climate.

In this combination, the crypto-climate is more important when the ability of the animal to respond is more limited, as by confinement in a pen without straw or baffles and without companions. When straw is present, an animal can establish its own micro-climate partly independently of the crypto-climate; the thermal relations are then much less predictable, and it is difficult to decide which environmental parameters should be measured for assessment of the micro-environment. It is probably not enough to devise instruments which record conditions close to the animal,

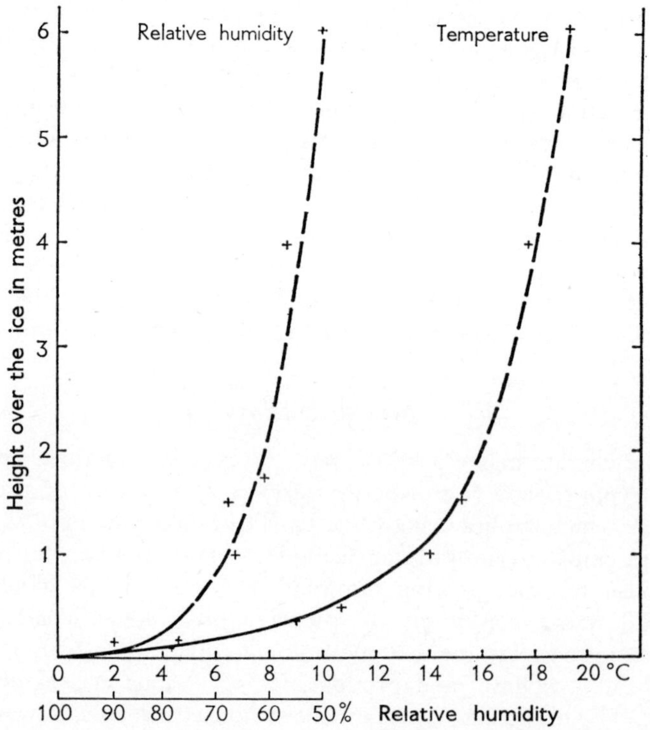

Fig. 1.3. Relative humidity and temperature of the air over the ice in a skating rink (from Hentschel, 1964, by permission of Elizabeth Licht).

because it is difficult to predict heat transfer rates from such measurements. The best integrator is probably the animal itself, in terms of the total heat production demanded of it by the environment for the regulation of its body temperature.

Special arrangements inside a building can produce marked variation in micro-climate over short distances. Fig. 1.3 shows how relative humidity and air temperature change rapidly with height above the ice in a skating rink. A very large part of the change occurs within the height of a man, so that the head of the skater is in an environment of about 15°C while his feet are in an environment close to 0°C—not a situation calculated to allow the highest efficiency in physical performance (Hentschel, 1964).

The great variation in micro-environmental conditions which occurs within very small distances is sharply illustrated by the choice of habitat of some poikilothermic species. An example of this occurs in the crypto-climate provided by the burrow of the desert woodlouse, *Hemilepistus reaumuri* (see Fig. 1.4). The animal's retreats constitute friendly micro-environments quite distinct from the hostile desert outside. The body temperature of

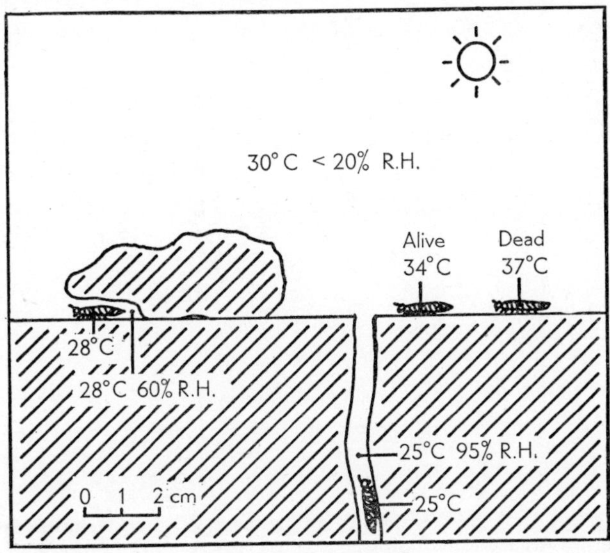

FIG. 1.4. Observations made in the habitat of the desert woodlouse *Hemilepistus reaumuri* (Isopoda), in the Algerian desert (from Edney, 1957, by permission of the University College of Rhodesia).

such animals is, of course, largely dependent on their physical environment, since, although this level may be raised by increased metabolism due to activity or other cause, it is not regulated by an internal mechanism like that operating in the homeotherm.

Homeotherms, however, also use environmental aids to thermoregulation. This is particularly true of certain small mammals which would find difficulty in surviving outside a relatively narrow temperature range. Hayward (1965) has made a study of deer mice, in which temperature-sensing thermistors attached to the animals

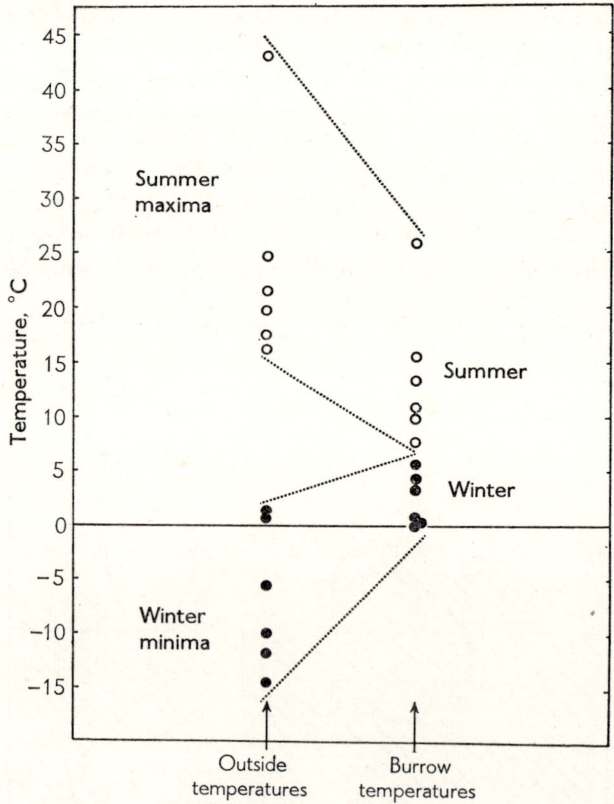

FIG. 1.5. A comparison of the seasonal outside-temperature extremes in each habitat with the average temperatures of the burrow of the deer mouse (*Peromyscus*) at the same times. The broken lines indicate the moderation of burrow temperatures in contrast to the seasonal extremes (from Hayward, 1965; reproduced by permission of the National Research Council of Canada from the *Canadian Journal of Zoology*, Vol. 43, pp. 341–50, 1965).

were pulled into their burrows. Fig. 1.5, taken from Hayward's paper, shows a comparison of outside and burrow temperatures in summer and winter, and provides evidence of the thermal stability offered by the burrow. Under desert conditions, for example, with a daytime temperature of 44°C, a mouse resting in its burrow at 26°C is not exposed to thermoregulatory stress. The thermal constancy of the burrow in contrast to the external 24-hour temperature variation is even more marked (Fig. 1.6).

The crypto-environment provided by an air-conditioned house is comparable to the mouse's burrow in providing relief

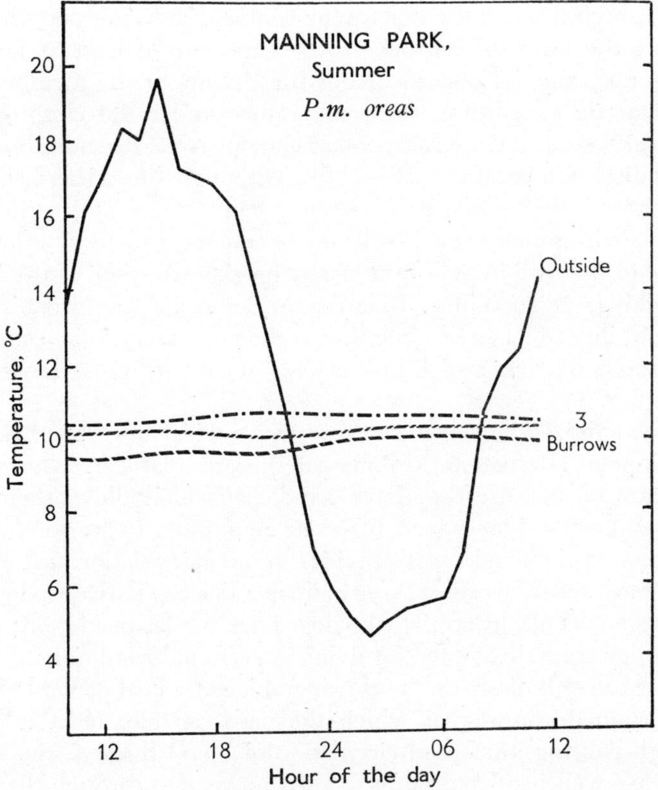

FIG. 1.6. A 24-hour recording of outside and burrow temperatures in the subalpine habitat of a deer mouse (*Peromyscus*) in summer, illustrating the thermal stability offered by the burrow (from Hayward, 1965; reproduced by permission of the National Research Council of Canada from the *Canadian Journal of Zoology*, Vol. 43, pp. 341–50, 1965).

from the demands of the external macro-environment. As Landsberg (1964) says, 'The house is man's primary invention to combat adverse climates'. This is true not only for himself but also for the animals he husbands, and in overcoming the effects of adverse climates the house is becoming an increasingly important tool in raising the efficiency of animal production.

SCALE OF ENVIRONMENT

The scale of the environment studied in relation to an animal is thus highly relevant: the macro-environment does not necessarily indicate the order of response to be expected, at least as far as energy exchange is concerned. For the Eskimo or the member of an Antarctic expedition, the temperature under the clothing is more relevant as an environmental factor than the surrounding air and radiant temperatures (Fig. 1.7). Although the micro-habitat may show little change as between a comfortable and a hostile macro-environment, the hostile environment provides a lower chance of survival in the event of any breakdown, such as the loss of clothing, or an injury. In order to determine the organism's susceptibility to its environment, it is thus necessary to investigate the various barriers which lie between it and its gross environment.

Some animals, however, can live over a wide range of physical environment (Barnett & Mount, 1967), particularly in respect of temperature, because they have a fur coat which allows them to withstand cold. Wolves have fur coats amounting to 60–70 mm in thickness on the trunk; this provides so much insulation that even in the most severe weather these animals can sleep curled up in the Arctic snow (Folk, 1966). Husky dogs have similar insulation; as a result they are well adapted for living in very cold conditions. Even so, they can still dissipate the considerable extra heat generated by the sustained running of which they are capable. Heat is lost through panting, increased circulation of blood through the skin of the less well-insulated limbs, and by movement through the air. Such an animal can adapt to a very wide range of environments, in consequence of its large body size, thick coat, and highly developed heat-loss mechanisms.

In the nineteenth century, Bergmann and Allen sought to establish general laws governing the geographical distribution of animals according to climate. Bergmann's Rule states essentially that, in a given species, subspecies of larger size will be found in polar regions, and of smaller size near the equator. Allen's Rule declares that the size of appendages in a given species decreases as

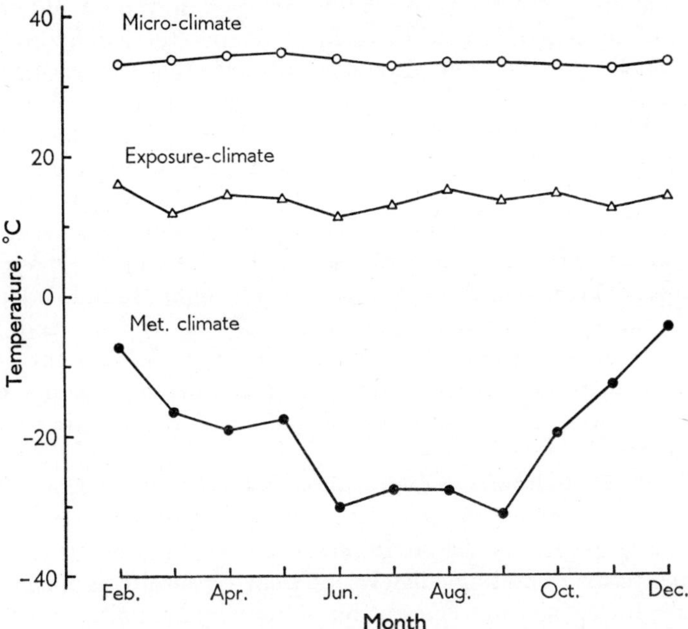

FIG. 1.7. Monthly mean values of the temperature of the meteorological climate, the exposure climate (within the station) and the micro-climate (under the clothing) to which four subjects were exposed during one year at an Antarctic station (from Norman, 1965; reproduced by permission from *British Antarctic Survey Bulletin*, No. 6, 1965, p. 7).

the mean environmental temperature decreases. These are generalizations, of course, with exceptions; the subject is discussed by Weiner (1966).

RESPONSES TO ENVIRONMENT

For experimental purposes it is often necessary to isolate components of the environment, so that, for example, temperature can be varied while other factors are held constant. Even if a component is not varied experimentally, there is still an implicit division of the environment into a given number of components when multivariate analysis is carried out. It is then necessary at some stage to integrate the components into the total environment, and to make measurements in the actual surroundings which have been the subject of analytical experiment.

An animal's environment may be analysed in terms of different sorts of sensory information, such as temperature, light, sound, and smell, and different physiologically effective factors such as high and low pressures, hypoxia, altitude, gravity, and acceleration. It is probably simpler, however, to divide environmental factors into two groups: those which are primarily thermal in their impact on the organism, and those which are not. The thermal factors include in addition to temperature (both air and mean radiant), air movement, humidity, and the physical nature of the floor or ground. The non-thermal factors include light, sound, odours, food, ionizing radiations, the nature of the terrain, and the presence or absence of other animals. Other animals are involved in social behaviour, reproduction, and reactions to predators, and under some conditions also make a thermal contribution as in huddling.

The total environment of an animal is highly complex, and it is probably impossible to factorize it completely in any useful way, since a large part of the environment cannot be characterized in simple terms. The immediately apparent physical components—temperature, air movement, humidity, light, noise, physical contacts—can certainly be measured very accurately, one by one, and by instruments separate from the animal. It is then possible to infer the *probable* impact which each of these makes on the animal, but the degree of certainty remains low except in predicting fairly gross effects such as that of temperature on metabolic rate in a highly standardized environment.

The stimulus to the animal to respond to its whole environment is more by way of being a compound error signal arising from the difference between the existing situation and the desired 'comfort' situation, than of being one of a number of individual physical variables which have been separated out for the purposes of analytical investigation. The variation of a single parameter at a time is an experimental situation and not a usual one. It is useful to remember this in the study of environmental physiology because of the over-riding importance attached by experimenters to measurable physical quantities, particularly temperature. It is relatively easy to get some measure of temperature, but its significance to the animal in a given situation is questionable. In a carefully controlled environment, a high correlation can be achieved between temperature and animal response, whether this is measured as food consumption, metabolic rate, shivering, or postural or activity adjustment. Such an environment is itself, however, an abstraction from reality. In the usual environment, temperature changes which give rise to clear-out effects in the laboratory may fade into insignificance in the face of other demands linked with the environment, such as the attacks of predators, competition and the search for food, and activity patterns associated with reproduction.

Direct application of the environmental physiologist's findings is therefore not always feasible. An increasing proportion of farm animals, however—in many countries now of the same order of number as the human population—live under conditions which restrict environmental variation. It is here that the application of knowledge may be expected to be beneficial.

2

SOME ASPECTS OF THE BIOLOGY
OF THE PIG

THE pig is now used increasingly in research. The considerable amount of research on the pig as a farm animal, particularly on its nutrition, which has been going on for some years, reflects the animal's great economic importance in meat production. In addition to this, the pig is coming to be used more and more in medical and physiological research, partly because in some respects there are physiological analogies between pig and man.

However, some of the similarities between pig and man may be more apparent than real. The relative lack of hair on the integument of each has led to the assumption that their skin is similar, but this is open to question. Although the skin of both contains similar proteins (Weinstein, 1966), and each has a sparse hair coat, a thick epidermis and a large content of elastic tissue, the dermis of the pig differs from man's in that it is poorly vascularized, and has only apocrine sweat glands; man's skin has a very rich blood supply and contains many eccrine glands (Montagna, 1966). A hairless pig has been described, and its skin examined histologically (David, 1932). The nerve supply to the skin of the pig has recently been investigated by Jenkinson and Blackburn (1967). The pig does not sweat in response to thermal stimuli (Ingram, 1967); indeed, the animal loses less water from its skin than does any other mammal which has so far been studied (see Chapter 6). Man, on the other hand, sweats abundantly in hot environments, and can lose more water in this way than any other mammal. In a number of other ways, including their higher growth rate and more acute sense of smell, pigs are manifestly different from man. Notwithstanding these differences, however, similarities of size and function are great enough to allow useful comparisons to be made between the two species.

ZOOLOGICAL FEATURES

The pig belongs to the order of mammals known as the Artio-dactyla, or even-toed ungulates, which contains three sub-orders: *Suiformes* (pigs, peccaries, and hippopotamuses), *Tylopoda* (camels), and the *Ruminantia*. Morris (1965) describes these in the following way. Within the *Suiformes* there is the infra-order of *Suina*, containing the super-family *Suoidae*, of which the family of *Suidae* constitutes the Old World Pigs (fossil pigs have been found in Europe and India, but not in America). This family has as its members:

Bush Pig	*Potamochoerus porcus*
Wild Boar (and domestic pig)	*Sus scrofa*
Pygmy Hog	*Sus salvanius*
Javan Pig	*Sus verrucosus*
Bornean Pig	*Sus barbatus*
Wart Hog	*Phacochoerus aethiopicus*
Giant Forest Hog	*Hylochoerus meinertzhageni*
Babirusa	*Babyrousa babyrussa*

Of these animals, *Sus scrofa* is the one with which this book is primarily concerned. Some classifications have previously sub-divided this species into a number of sub-species. Thus the Indian pig was termed *Sus cristatus*, the Chinese pig *Sus indicus*, and the European pig *Sus scrofa*, while the domestic pig became *Sus scrofa domesticus*. It is doubtful, however, whether such sub-divisions are either useful, or valid on zoological grounds. The true origins of the pig are probably lost in obscurity.

Sus scrofa is distributed over the woodlands of Europe, and in North Africa and Asia, while island races are to be found in Ceylon, Sumatra, Java, Japan, and Formosa. The wild boar was once very plentiful in Europe, where it has been hunted for centuries, but it is now restricted to a few areas of woodland. It has long been extinct in Britain. Crandall (1964) points out that although pigs show wide variation in size, colour, and shape of tusks, they all have in common the flat rounded snout on a flexible muzzle. The upper canine teeth turn upwards to form tusks, and sometimes are very large, particularly in males. The animal has four toes on each foot, with only the third and fourth functional. The hair is coarse and sparse. The pig can be described as omnivorous, although vegetable material forms the bulk of its food. The gestation period of *Sus scrofa* (wild pig) is quoted as ranging between 101 and 130 days, an average of 116 days, as compared with 112–15 days for domestic pigs for various breeds.

The features of the environment which elicit the greatest physiological and behavioural responses of an animal vary from mammal to mammal, and vary also with the age of individual mammals. This generalization is illustrated in the domestic pig by the susceptibility to cold of the new-born pig, and the susceptibility to heat of the large mature domestic boar or sow. In its natural condition, however, the pig lives in woodland, where it can cool itself by wallowing in mud, so that the use of exogenously derived water compensates for the animal's failure to sweat effectively (see Chapter 6). Adult pigs will die when exposed to air temperatures over 36°C under sunny conditions, but given access to mud wallows they can withstand exposure to this temperature indefinitely. In its natural habitat, therefore, the large animal overcomes by a behavioural adaptation its inability to cope by sweating with hot conditions. Similarly, the inability of the new-born pig to combat cold is overcome by the farrowing and nursing behaviour of the sow, which makes a nest for the piglets in which they can remain warm and dry, and to which they can return after brief forays outside.

Some aspects of the history of the domestication of pigs show how the animal's natural relation to its environment has been affected during the course of development of pig husbandry.

HISTORICAL

Intensive selection of pigs for domestic purposes began in the eighteenth and nineteenth centuries, with the result that there are now many distinctive breeds which are used for meat production and vary according to local preference and market requirements. In Britain, the Large White and Landrace are now the most common breeds husbanded, with some coloured pigs such as the Wessex Saddleback and the Berkshire. The Large White is known in many parts of the world as the Yorkshire pig, and has played a large part in the development of the domestic pig. Its origin is commonly said to derive from the selection efforts of one, Joseph Tuley, a Yorkshire weaver. The Duroc pig in the United States has reached the greatest numbers in total production. General descriptions and photographs of the most common breeds are to be found in Davidson's (1954) book on the production and marketing of pigs, and in the work on pig diseases by Anthony & Lewis (1961).

The greatest part of the world pig population now consists of the domestic pig (see Table 2.1). The estimated world population of pigs is in excess of 500 millions, but this is only half the total

TABLE 2.1

Numbers of pigs in the world, in millions (from Food and Agriculture Organization Production Yearbook, 1964)

	Year	
	1947–52	*1962–3*
Europe	69·4	109·2
U.S.S.R.	19·7	69·9
North and Central America	76·0	82·8
South America	35·3	65·0
Asia	21·3	38·9
China	73·8*	—
Africa	4·1	5·3
Oceania	1·9	2·4
World	295·7	553·5

* Average of three years.

number of either sheep or cattle (Food and Agriculture Organization, United Nations). In relatively primitive peasant economies the pig is still generally free to seek its own environment to a large degree, that is, to wander through settlements and adjacent areas; the animals are removed for human consumption as required. In medieval times in Great Britain, for example, pigs obtained a large part of their food supply from woodland; they were collected in the morning by communal swineherds, who would set their charges to graze on pastures or search for other food, and then return the pigs to sties at night. An indication of the early economic significance of pigs is given in the Domesday Book, where areas of land are sometimes described not in area measure but in terms of 'keep for swine' (Trow-Smith, 1957). Pigs are still herded today in some woods in Europe (Davidson, 1954).

Following the agricultural land enclosures in Britain from 1700 to 1845, pig husbandry became more intensive owing to the diminished land available, and the tendency was to confine the animals to their sties. Davidson (1954) points out that these small buildings, from four to seven feet square and with an open yard of about the same size, were excellent for providing night-time shelter for pigs which had been out all day, and which had little fat or other insulation for protection from wind and rain while resting. However, when the animals were confined in small spaces during the day as well, and deprived of their natural foraging activity, they began to form considerable subcutaneous fat. The modern practice is to confine pigs in groups in houses of various types, so that now in travelling across Denmark, a major pig-producing country, one may not see a single pig out of doors. The emphasis is on the selection of animals which do not become excessively fat under conditions of being permanently housed, and which will respond to controlled levels of food intake by the production of muscle and little fat. What is being attempted is to obtain by selection a result which the physiological stimulus of exercise would otherwise have provided, that is muscle hypertrophy.

It is not surprising, therefore, that the present domestic pig, the result of intensive breed selection, in its unexercised and confined state, with rapid early growth and later restriction of food intake, and with the denial of many of its inherent behavioural responses, should differ from the wild pig in its pattern of overall response to its environment. The wild pig is vigorous, active, and aggressive; the domestic pig is a very different creature, content largely to feed and sleep, inclined to be obstreperous only if disturbed. Pig husbandry for meat production is carried out increasingly under conditions of controlled environment, so that physiological findings can be applied in practice more readily than is the case for many other types of livestock. In addition, the animal when young has relatively little thermal insulation, and presents during the course of growth a considerable range of body size; both factors make the animal useful in the study of thermal environment and animal energetics.

MINIATURE PIGS

The pig is inquisitive, and in some respects difficult to handle, particularly when large. Strains of 'miniature pigs' have been developed in an attempt to overcome this difficulty in the laboratory (Weaver & McKean, 1965; Bustad, Horstman & England,

1966). Such pigs approach the order of size of the naturally occurring small pigs of Indonesia, Bhutan, Nepal, and Vietnam. In the United States, the Hormel and Pitman-Moore strains of miniature pig have been developed by selection from wild pigs descended from those introduced by the Spaniards in the sixteenth century. They have an average birth weight rather less than 1 kg (in the standard domestic pig the corresponding weight is somewhat higher than 1 kg); at six months old they weigh about 20 kg (90 kg for the standard pig). Miniature pigs for biomedical research have been produced successfully in Germany from foundation stock made up from American miniature pigs, Vietnamese pigs, and the German Improved Landrace. Continued back-crossing to the miniature pig has established a combination of small size with the white colour and high fertility of the Landrace. At 5 months of age the animals are about 17 kg in weight. The first evidence of oestrus can be seen in the gilts soon after 9 weeks of age, and they are sexually mature at $4\frac{1}{2}$ to 6 months (about 8 months for the standard pig); the boars can be used for breeding at 3 to 4 months of age (Haring, Gruhn, Smidt & Scheven, 1966). In addition to their similarities to standard pigs, and their consequent value in research on the pig itself, miniature pigs represent a real contribution to the available choice of laboratory animals (Kaemmerer, 1966).

PHASES OF DEVELOPMENT

During the course of its development from the newly born to the mature animal, the pig shows at least three phases in its responses to environmental stimuli. The new-born phase, which is taken here to cover the first week of extra-uterine life, is the first of these; during this phase the animal is particularly sensitive to cold, and shows certain peculiarities accompanying post-natal adaptation (see Chapter 4). The pig weighs rather more than 1 kg at birth, and is a member of a litter which is usually six to twelve in number; post-natal growth is rapid, so that after one week the animal has approximately doubled its birth weight.

The second phase is that of the growing pig; it includes weaning, which usually takes place at five to eight weeks, and lasts until about seven months of age. This second phase is marked by a rapid increase in size which, by itself, without any other change being necessary, diminishes the animal's cold-sensitivity progressively. Vigorous muscular activity normally occurs during this second phase, if it is allowed by the husbandry system adopted; if this activity is prevented, fat deposition increases in the latter part of the period. The chief economic problem relating to the domestic pig arises here: to achieve rapid growth in muscle, while keeping exercise to a minimum in order to reduce costs of feeding. The third phase of the pig's development is marked by sexual maturity and reproductive activity. The animal is by now relatively insusceptible to cold, in consequence of its large size and increased tissue insulation, but is highly susceptible to heat, ambient temperatures in excess of 30°C proving progressively more embarrassing, and requiring wallows, mud baths, or water-sprays to provide evaporative heat loss from the animal's surface if hyperthermia is to be prevented.

Pigs characteristic of each of these three phases of development differ from each other particularly in respect of their overall thermal insulation. In the first phase, the new-born, insulation is minimal, and cold is resisted to a large degree by the behavioural adaptation of huddling with litter-mates and with the sow. In the third phase, by contrast, thermal insulation is high enough for pigs to live in the Arctic with apparently little discomfort (Irving, 1956*a*, *b*). The new-born pig brings out in sharp relief, and quantitatively to a greater degree than the older pig, those features which characterize the relation between the animal and its thermal environment.

It is probable that some of the confusion which arises in applying experimental results to the husbandry situation does so on account of failure to recognize the changing significance of a given environment during the pig's development. The growing lamb, for example does not present the same problem, nor does the calf; at birth these animals already have a high thermal insulation (Alexander, 1961*b*; Roy, Huffman & Reineke, 1957) and their ability to withstand cool conditions is obvious from their mode of life. The outstanding example of this lies in the cold resistance of the reindeer calf, *Rangifer tarandus*, which is born in the snow and is following the herd, with its mother, within minutes of birth. The pig, however, is much more akin to man in its thermal relations than it is to these other species, although the economic importance of the pig for agriculture has put the animal into the hands of those whose knowledge and experience are naturally founded on the behaviour of other farm animals, rather than on that of man.

Before proceeding to a closer examination of the effects of

environment on the pig, it is appropriate to pay some attention to the foetus and new-born. Metabolism and temperature are then considered in Chapter 3; following this, these topics are discussed in relation to the new-born and growing pig in Chapters 4 and 5 respectively.

THE FOETAL AND NEW-BORN PIG

Perry (1956) made a study, over a period of four years, of 247 pregnancies in 85 sows and gilts in one herd of Large White pigs. He found the mean gestation period to be 114·2 (±0·14) days, with a range varying from 104 to 126 days. There was no relation between litter size and extremely long or short gestation, and no difference in the average duration of pregnancy as between young and old sows. The mean number of pigs born (both live and dead) was 13·1, and ranged from 3 to 24. The mean number born alive

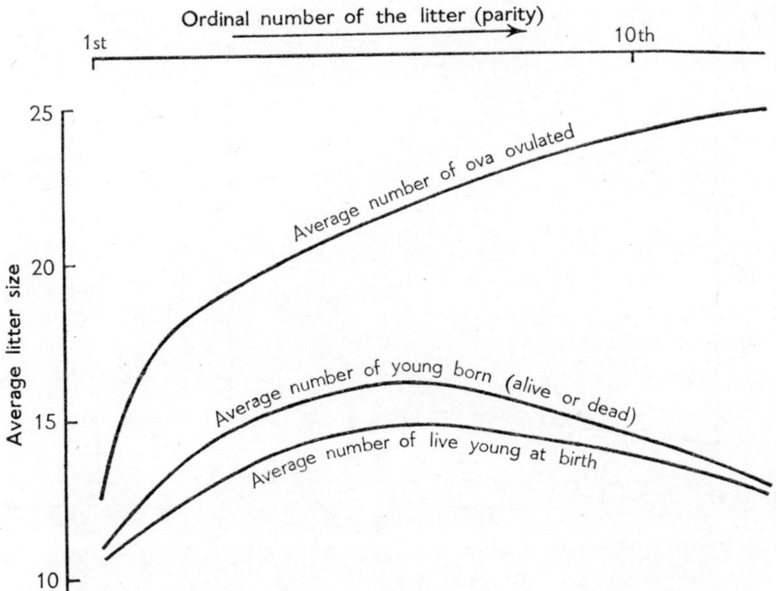

FIG. 2.1. The approximate relation between average litter size, embryonic mortality, the incidence of stillbirths, and the ordinal number of the pig in the litter (from Perry, 1956, by permission of *Journal of Agricultural Science*).

was 12·4, and the mean number weaned was 8·1. In this connection, a survey of piglet deaths in England and Wales (Veterinary Investigation Service, 1960) established that 17% of piglets born died by three weeks of age. The number of pigs born rose with successive pregnancies to a maximum of more than 14 in the fifth litter, and thereafter declined; embryonic mortality increased progressively. These trends are summarized in Fig. 2.1.

The anatomical and physiological development of an animal at birth can be modified considerably by circumstances occurring during pregnancy, so that animals of the same gestation length do not at birth necessarily show the same levels of physiological or anatomical development (Hammond, 1961). The practical importance of this is that the larger and more functionally developed a pig is at birth, the more likely is it to survive and grow satisfactorily in an environment in which it might otherwise die. The changes

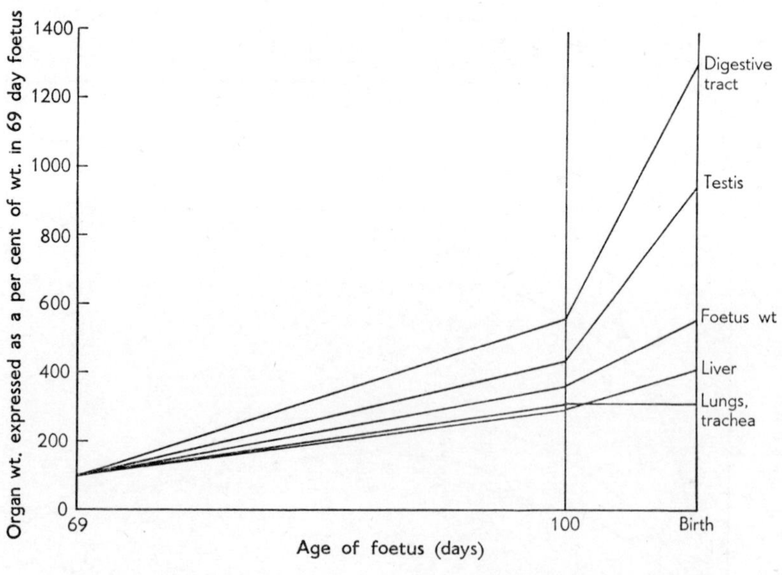

FIG. 2.2. Developmental changes in organ weight in the prenatal life of the pig. The weights of organs at 100 days post-coitum and at birth are expressed as a percentage of those at 69 days post-coitum (from Hafez, Mauer & Ensminger, 1958, by permission of *Growth*).

found in organ weight in pre-natal life are given in Fig. 2.2, which shows the greater relative growth in the digestive tract and testes as compared with liver and lungs.

THE FOETUS

Nutrition of the foetus, and space in the uterus, are the main factors related to size at birth. Three main nutritional phases occur during pregnancy; the blastocyst, the embryonic, and the foetal. During the blastocyst stage, nutritional requirements are met by diffusion from the uterine secretions. In the embryonic stage, the trophoblast determines the food supply to the embryo regardless of the nutritional state of the mother. Following the atrophy of the trophoblast cells, which occurs at about the 70th day of pregnancy in the pig, the placenta no longer increases in area, and the foetal stage is entered. In the foetal stage of some mammals, nutrition depends on the level of nutrition of the mother, and high-plane nutrition during this stage tends to increase the size of the newly born. This was found to be the case with lambs, where pre-natal development was markedly affected by the level of nutrition of the ewe during the latter part of pregnancy (Wallace, 1948). In the pig, however, Hafez, Mauer & Ensminger (1958) found no significant effect on the length or thickness of bone in the foetus due to plane of nutrition, and no effect on organ weights. From experiments in which sows were fed at different planes of nutrition during three pregnancies, Elsley, Bannerman, Bathurst, Bracewell, Cunningham, Dodsworth, England, Forbes & Laird (1967) found that the total litter-weight at birth was unaffected by the level of feeding during pregnancy.

Pomeroy (1960*d*) considers that the situation in a polytocous species like the pig may be peculiarly different from that in a monotocous or ditocous species like the sheep; in both there is nutritional competition between the tissues of mother and foetus, but in the pig there is competition between the foetuses themselves. In this respect the smallest pig in a litter is typical of an animal on a submaintenance diet, that is to say it has disproportionately little muscle. The consequences of this are that the small pig's metabolic capability and cold-resistance (see Chapter 4) are low, and it is not a strong competitor for a teat.

Embryonic mortality, which is the chief factor determining the eventual size of the litter born to the sow, is the result of many factors (Rathnasabapathy, Lasley & Mayer, 1956; Perry & Rowlands, 1962). One of the most important of these, probably, is that space in the uterus can be limiting, with overcrowding contributing to mortality. Infertility and embryonic and neonatal mortality in the sow have been discussed by Pomeroy (1960*a*, *b*, *c*, *d*); in an analysis of pre-weaning mortality, he found that 70% of all deaths occurred in the first three days after birth, and that the mean birth weight of pigs which died within three days was only 1·00 kg as compared with 1·26 kg for those which survived. Mortality was highest in litters of less than 5 and more than 15 in number, and showed a marked seasonal variation, being highest in the winter.

In the post-natal period, environmental temperature and other factors important in determining the new-born pig's heat loss play a large part in determining the animal's survival (see Chapter 4). The foetus, however, lives in the constant temperature environment provided by the mother's uterus. Probably the major part of its metabolic heat production is dissipated in the placenta, where counter-current heat exchange is likely to be very efficient.

Allen and Lasley (1954) have investigated the influence of the season of farrowing on the performance of 172 sows and 1,635 pigs in Missouri. More and heavier pigs were born in the autumn than in the spring, but the weight difference disappeared by the time the pigs were 5 months old. Survival rate and growth rate both progressively decreased in succeeding litters during the autumn.

GROWTH AND BODY COMPOSITION IN THE NEW-BORN PIG

The pig at birth weighs usually rather more than 1 kg, and the animal grows rapidly. Associated with this rapid growth is the availability of milk from the sow; milk can be expressed from the sow's udder usually one to two days before farrowing. The pig doubles its birth weight in one week, approximately, whereas man takes about six months. In this respect, man is the unusual organism, as reference to Fig. 2.3 shows; the slow pre-pubertal gain in weight in man has no counterpart in the other animals represented.

The fat content of the new-born pig is about 1%, a value obtained by Widdowson (1950) and confirmed by McCance & Widdowson (1959). Such low values are very different from the mean value of 16% fat found in the human infant (Widdowson, 1950). Wood and Groves (1965) have followed in detail changes in

the pig's weight and body composition from birth. They found
that at an age of about 18 days and at a body weight of 5 to 6 kg
there was a temporary slowing of the growth rate; they suggest
that at this point the increase in the sow's rate of milk production
is no longer able to keep pace with the full demands of the litter,

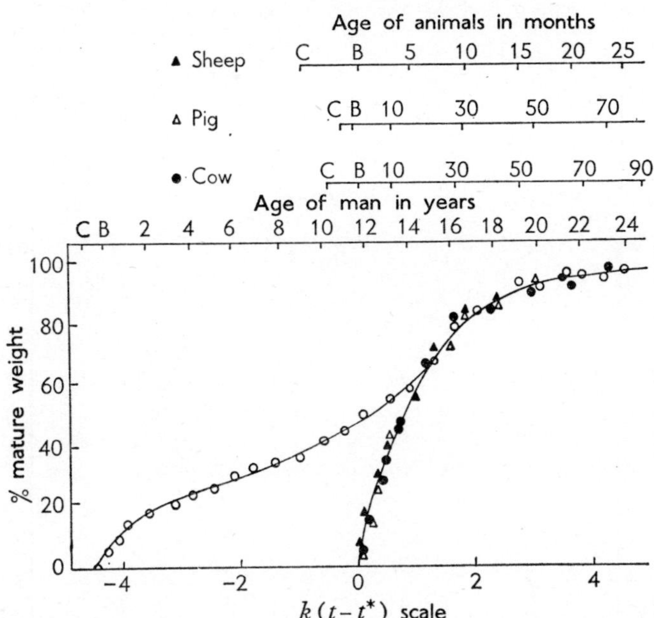

FIG. 2.3. The weight-growth equivalence of animals and man.
The continuous line originating at zero on the abscissa is the mean
value for cow, pig, sheep, rabbit, fowl, rat, mouse, guinea-pig,
pigeon, and dove; the line with open circles refers to man. The zero
of the abscissa is defined by $t = t^*$, where t^* is the intercept derived
from logarithmic extrapolation. The curves emphasize the slow
pre-pubertal growth in man as compared with the animals represented (from Samuel Brody, *Bioenergetics and Growth*, Reinhold
Publishing Corporation, New York, 1945).

and that the young pigs are not yet ready to make full use of the
additional feeding stuffs ('creep feed') normally provided for them
under the conditions of intensive husbandry. It is well known that
the sow's milk production reaches a maximum at three to four
weeks after farrowing; when the piglets increase their consumption of creep feed there is a corresponding decrease in the amount
of sow's milk required to produce unit gain in weight of the pigs

(Fig. 2.4). The percentage of body water in the piglet declines from 83% to 67% in the first fifteen days after birth, a decline which is characteristic of the new-born mammal. Wood and Groves attribute this to a rapid increase in body fat and a slower increase in protein, but body protein is increasing more rapidly than body water during this period.

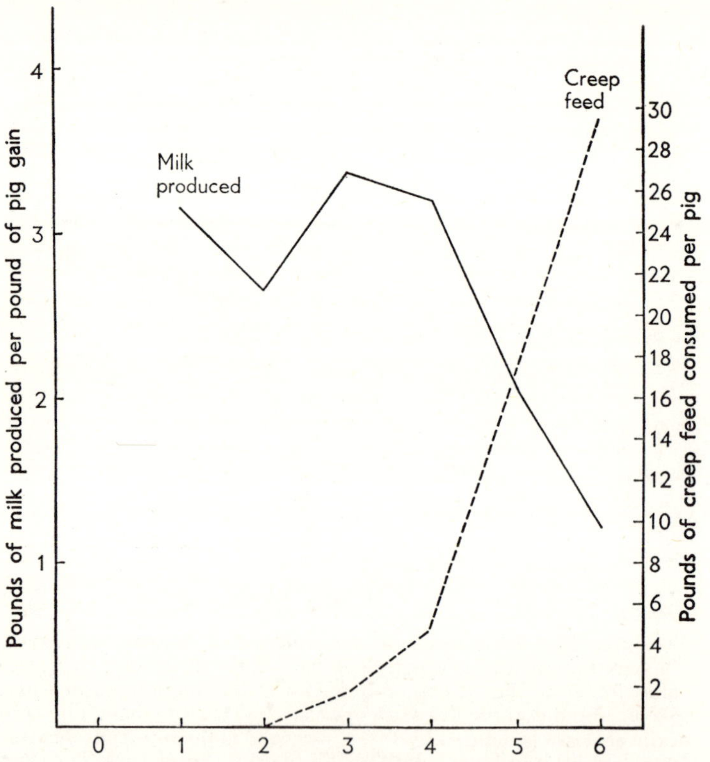

FIG. 2.4. The sow's milk production per unit gain of weight in the pigs of her litter, and the creep feed consumption of the pigs (from Allen, Lasley & Tribble, 1959, by permission of the Agricultural Editor, University of Missouri).

The increase in body protein content can be accelerated by feeding a diet which is 50% protein from several days of age (Filer & Churella, 1963). Such pigs gain weight more rapidly than litter-mates on a 14% protein diet, and show hypertrophy of skeletal muscle and earlier maturation in respect of body composition; they also show a lower fat content (see Fig. 5.24).

FOOD REQUIREMENTS

Although initially the new-born pig is dependent on the sow's milk not only nutritionally but also for the passive immunity it derives from colostrum, after the first few days the animals can be reared by artificial feeding. Knowledge on this subject has been summarized in an Agricultural Research Council (1967) publication. In the ordinary way, however, the pigs are almost entirely dependent on milk for nutrition for the first few weeks, after which they take increasing quantities of supplemental feeds.

When piglets feed from the sow, three stages of activity are usually described. In the first stage, the piglets nose the udder vigorously when the sow lies down in position for suckling. In this phase, the liberation of oxytocin from the posterior pituitary leads to 'let down' of milk in the mammary gland; the first stage usually lasts for one to two minutes. In the second stage the piglets become quiet, and it is in this stage, which lasts only about 20 seconds, that they obtain milk. In the third stage, the piglets once again begin to nose the udder, and the sow may rise; this marks the end of the suckling period.

Barber, Braude & Mitchell (1955) reviewed the literature on the frequency of suckling under natural conditions, and found the average interval between successive sucklings to be about 1 to $1\frac{1}{4}$ hours. From their own measurements, they found that both the quantity of milk obtained by the young pigs, and their gain in body weight, were greater when suckling was allowed every hour than when it was restricted to about every three hours. They found the average eight-week lactation yield from a sow (Large White) to be about 350 kg; this is a higher value than that obtained by other workers. Allen, Lasley & Tribble (1959) give figures for the milk production of sows, taken from various authors as well as from their own studies. They found that Landrace sows were highly efficient in converting feed to milk, and their young pigs were very efficient in converting feed to weight gains. Duroc sows, however, were not so efficient in milk production, although Duroc piglets were as efficient as Landrace in weight gain. The average milk production for a six-week lactation period was 139 kg for the Landrace and 107 kg for the Duroc sow. Fig. 2.5 illustrates lactation curves of sows; the results show the maximum milk yield occurring characteristically at about three weeks from the onset of lactation.

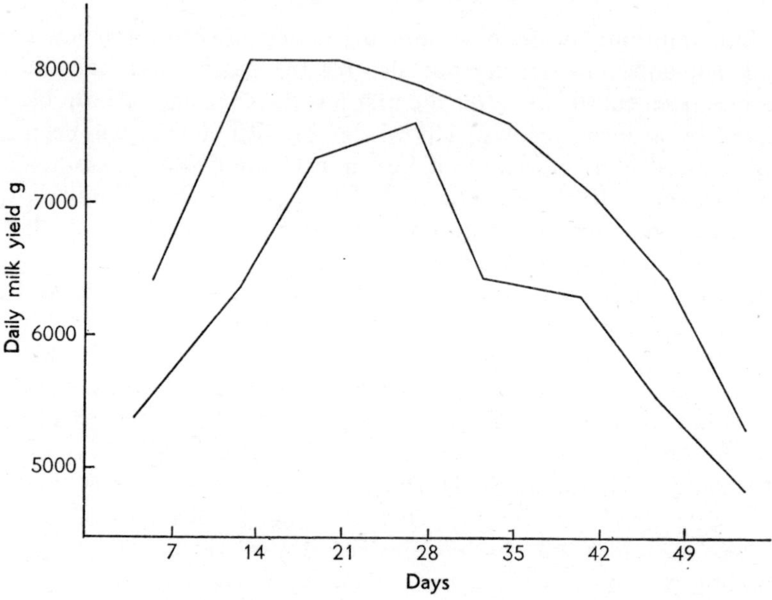

Fig. 2.5. Lactation curves of sows (from Braude, 1954, by permission of Butterworths Publications Ltd.).

Fig. 2.6 shows the effect of the stage of lactation on the composition of the milk, the decline in total solid content during the first two weeks particularly, and the simultaneous increase in fat content. Mean values for the composition of sow's milk are given in Table 2.2. There is a sharp decline in protein content between the colostrum of the first day of lactation and the milk of the third

TABLE 2.2

The mean composition of colostrum (1st day milk), and milk later in the sow's lactation (from Bowland, 1966)

	Stage of lactation		
	1st day	3rd day	mean: 1–8 weeks
Total solids %	22·8	24·3	20·1
Fat %	6·6	12·0	6·8
Protein %	12·8	8·6	7·3
Lactose %	3·2	4·3	5·1
Ash %	0·73	0·89	0·99

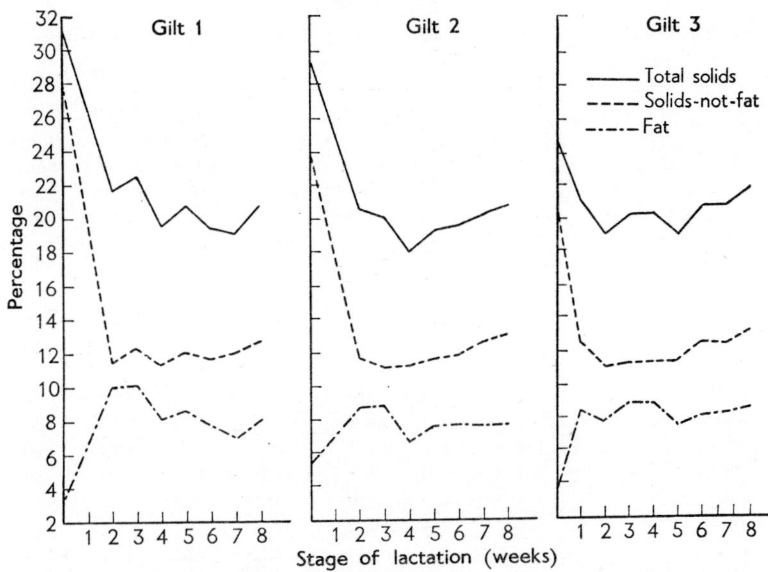

FIG. 2.6. The effect of the stage of lactation on the composition of milk from three gilts (from Barber, Braude & Mitchell, 1955, by permission of *Journal of Agricultural Science*).

and subsequent days; the fat content rises during lactation, then subsequently falls. Further details of the analysis are provided by Bowland (1966). In Table 2.3 the composition and equivalent energy content of sow's milk are compared with those of the milk of other species.

The consumption of milk by the piglet can be measured by the 'test-feeding' long practised on the human infant, a method which is open to errors, particularly the loss of weight which occurs during the suckling period, between weighings. This has been investigated by Spaendonck & Vanschoubroek (1964), who found the mean weight loss, which they ascribed to metabolic processes, to be about one gram in a two-minute period. If the sow produces about 3 kg of milk per day, and gives 24 feeds to 10 pigs, each pig obtains about 12 g of milk at each feed. An error of 1 g thus approaches the 10% level, and cannot be ignored. After about two weeks of age, the milk intake of each pig at each feed is of the order of 30 g (see Table 11.1).

The artificial feeding of baby pigs has prompted a considerable

TABLE 2.3

The composition and energy content of milk from various animals (from Kleiber, 1961, by permission of John Wiley & Sons, Inc.)

Source of milk	Composition			Energy in components per kg of milk			Energy per kg of milk (kcal)
	Fat %	Lac-tose %	Pro-tein %	Fat % ×92·0 (kcal)	Lac-tose % ×39·5 (kcal)	Pro-tein % ×58·6 (kcal)	
Rat	9·3	3·7	8·7	856	146	510	1512
Woman	3·8	7·0	1·2	350	276	70	696
Sow	7·0	4·0	6·0	644	158	352	1154
Cow	3·7	4·8	3·3	340	190	193	723
Sheep	6·2	4·3	5·4	570	170	316	1056
Deer	10·5	4·5	9·0	966	178	527	1671
Goat	4·1	4·7	3·3	377	186	193	756
Mare	1·7	6·6	2·2	156	261	129	546
Bison	1·8	4·6	4·0	166	182	234	582
Reindeer	22·5	2·5	10·3	2070	99	604	2773
Musk ox	11·0	3·6	5·3	1012	142	311	1465
Water buffalo	12·0	4·0	6·0	1104	158	352	1614
Rhinoceros	0·3	7·2	3·2	28	284	188	500
Elephant (African)	20·5	7·3	3·2	1886	288	188	2362
Fin whale	32·0	0·3	13·0	2944	12	762	3718
Blue whale	42·0	1·0	12·0	3864	40	703	4607
Porpoise	49·0	1·3	11·0	4508	51	645	5204

Milk constituents: lactose, 3·95 kcal/g, casein, 5·86 kcal/g, butter fat, 9·20 kcal/g.

amount of work on suitable components of the diet and on the pig's gastrointestinal tract, since the animal will eat solid food as early as one to two days of age. On account of its palatibility, sucrose might be considered a useful constituent of the diet. Becker, Ullrey, Terrill & Notzold (1954), however, found that a diet containing 56·6% sucrose caused diarrhoea and a high mortality when given to pigs 1 to 10 days old; replacing the sucrose by glucose produced satisfactory results. The sucrose diet was more satisfactory in older pigs. These observations accorded with the later findings of the absence, or the presence in only very low concentrations, of the enzyme sucrase, in the intestinal mucosa of the new-born pig (Walker, 1959; Dahlqvist, 1961). From their experiments on the digestion of sucrose by the piglet, Kidder, Manners & McCrea (1963) considered that the mucosa of the

6-day-old pig has little capacity to hydrolyse sucrose; they determined this by measuring the rise in plasma fructose after feeding sucrose in the diet. By 17 days of age the capacity to hydrolyse sucrose had increased, so that equivalent quantities of sucrose or invert sugar in the diet raised the plasma fructose to a corresponding extent. From experiments in which new-born pigs were injected intravenously with fructose, Kidder, Manners, McCrea & Weaver (1963) found that the metabolism and excretion of fructose were very slow, but the increased disappearance of fructose in older pigs suggested the development of a metabolic ability which increased with age.

HYPOGLYCAEMIA IN THE NEW-BORN PIG

Following a study of piglet mortality in Illinois, Graham, Sampson & Hester (1941) published a description of a condition given the rather non-specific name of 'baby pig disease'. Pigs which had been normal at birth showed at 24 to 48 hours of age signs of weakness and loss of appetite; the animals left the nest, lapsed into coma, and died. No gross pathological lesions were found, and no infectious agent. The only abnormal finding was a mean blood glucose level of only 26 mg/100 ml. which was very low when compared with the normal value (Shaffer–Hartmann–Somogyi method) for pigs at this age of 113 mg/100 ml. In the early stage of symptoms, injections of dextrose and the forced feeding of milk prolonged the life of the affected pigs, and in some cases the pigs so treated recovered.

When spontaneous hypoglycaemia of this kind had been recognized, Hanawalt & Sampson (1947) investigated the relation between age and time of onset of the acute hypoglycaemia brought on by fasting in new-born pigs. They found the animals to be highly susceptible to acute hypoglycaemia both immediately after birth, and during the rest of the first week, if food intake was markedly restricted. By the time the pigs were a week old, however, it became more difficult to produce a fatal hypoglycaemia by fasting (Fig. 2.7), and the pattern began to move towards that characteristic of the fasted adult animal. Glucose was found to be the most effective sugar capable of resuscitating pigs in hypoglycaemic coma; mannose had some effect, but fructose had none (Newton & Sampson, 1951).

This work was succeeded by an investigation by Morrill

(1952*a*, *b*, *c*, *d*) into baby pig mortality with particular reference to hypoglycaemia. Morrill confirmed that the syndrome produced by fasting resembled clinically the naturally occurring 'baby pig disease', and in these situations the hepatic glycogen was rapidly depleted and severe and fatal hypoglycaemia rapidly developed. The rate of development of hypoglycaemia depended on the environmental temperature; at 15°C the animals were moribund

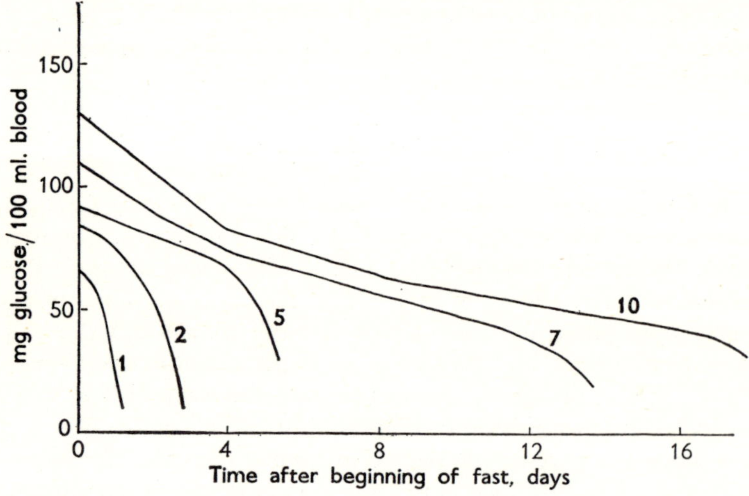

FIG. 2.7. The effect of fasting on the blood glucose level of the new-born pig. The figures on the curves give the age in days at the beginning of the fast; the older the pig at the onset of fasting, the more difficult is it to produce hypoglycaemia (constructed from results of Hanawalt & Sampson, 1947).

in 28 hr, but at 31°C it took 84 hr for the condition to set in. This suggested that the reduction in the animal's heat loss due to raising the environmental temperature conserved the carbohydrate reserves. The pigs in the warm lost 31% of their body weight before death, while pigs in the cold lost only 12%; it is possible that the warm pigs could use metabolizable reserves denied to the cold pigs. This is suggested also by the final weights of the livers: 16 g in pigs kept in the warm, and 23 g in the pigs in the cold. The smaller percentage loss in body weight in pigs kept in the cold was confirmed by Kemény, Gáspár, Pethes, Tóth & László (1955), as was the prolongation of survival time in warm conditions. Goodwin (1957*a*) showed that reduction of the new-born pig's blood glucose concentration produces a syndrome of complete collapse which is indistinguishable from the state during starvation, and from this he concluded that the metabolism of the new-born pig is governed completely by the concentration of circulating glucose; heart and respiratory rates, and body temperature, all declined with the glucose level.

Some of the problems raised by the apparent dependence on glucose were investigated by McCance & Widdowson (1959), when they kept fasted new-born pigs at 12 and 31°C environmental

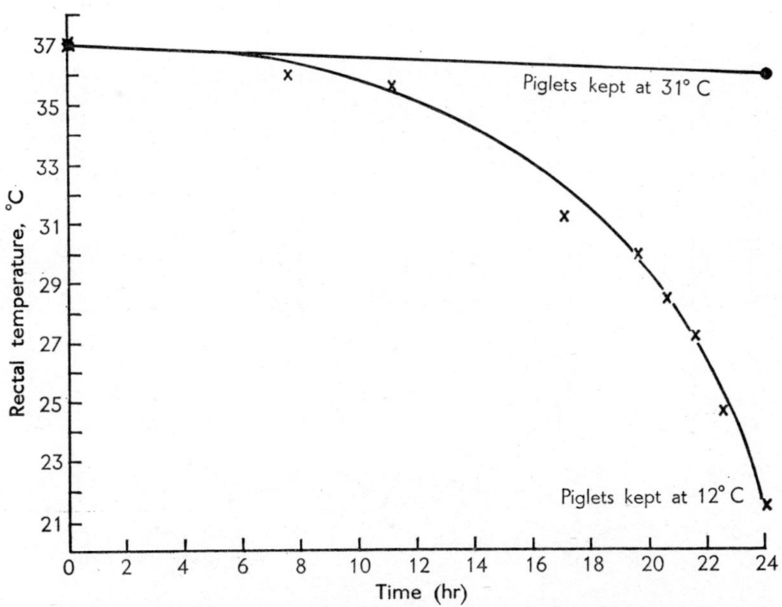

FIG. 2.8. The average rectal temperatures of fasted new-born pigs in one litter; four piglets were kept at 12°C and four at 31°C for 24 hr (from McCance & Widdowson, 1959, by permission of *Journal of Physiology*).

temperature and made carcass analyses. Fig. 2.8 shows the time-course of rectal temperature in the two groups. Cold intensified the rate of metabolism but made no great difference to the pattern of the materials being metabolized. The amount of fat in the carcass was small (1%), and could therefore make only a small contribution to the metabolic requirements. The total solids metabolized at 12°C were 35 g/kg.24 hr, and at 31°C were 17 g/kg.24 hr. It was calculated that the glycogen which disappeared between birth and death was 69 and 67% of the total solids metabolized in the animals kept cold and warm respectively. The actual figures for glycogen consumption were 24 g/kg.24 hr. at 12°C and 11·2 g/kg.24 hr at 31°C, and although the 12°C animals catabolized more tissue protein than the 31°C pigs, it was clear that carbohydrate provided the main source of energy for the

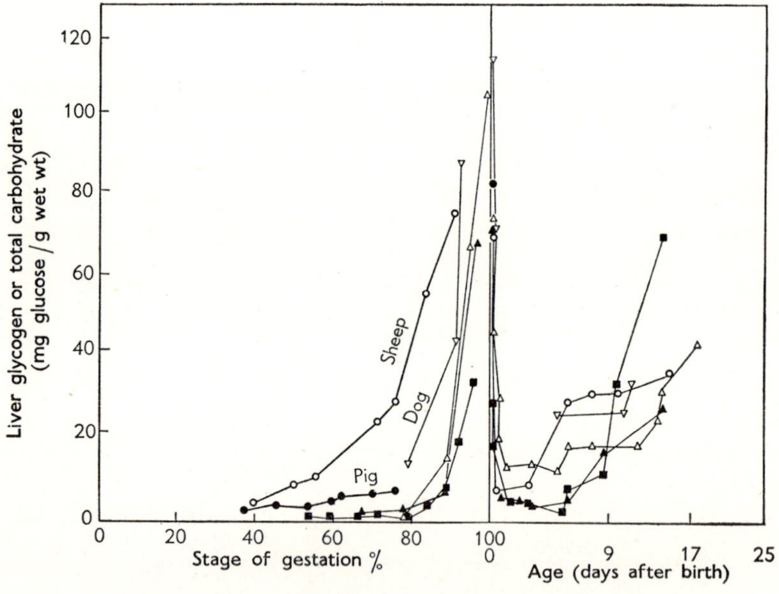

FIG. 2.9. The concentrations of liver glycogen in different species before and after birth. ▲ rabbit; △ rat; ■ guinea-pig (from Shelley, 1961, by permission of *British Medical Bulletin*).

fasted new-born pig. Like other mammals, the pig is born with a high concentration of glycogen in its liver (see Fig. 2.9), and its skeletal muscle contains particularly large quantities of glycogen (Shelley, 1961; Dawes & Shelley, 1968).

McCance & Widdowson found that the cold pigs contained more water than the warm animals (confirmed by Elneil & McCance, 1965), and the dependent parts of the animals were marked by the pattern of the floor of the cage in which they were lying. They noted in the 12°C pigs that, especially towards the end of the experiment, the animals felt very cold to the touch, although they had a pink bloom and did not appear cyanosed. The human infant presents a similar appearance in the condition of 'neonatal cold injury' (Mann & Elliott, 1957), which can occur in inadequately heated houses during cold weather, as well as in more extreme conditions. In the first week or two of the baby's life, what is usually seen in this condition is increasing apathy and the rejection of food. The body temperature is nearly always well below 32°C. The only important biochemical abnormality found by Mann & Elliott was lowering of the blood glucose level, sometimes down to 10 mg/100 ml. or less. Glucose administration and slow re-warming were very effective in the treatment of these infants.

The new-born calf and foal, unlike the new-born pig, do not readily become hypoglycaemic during starvation (Goodwin, 1957*b*). The new-born lamb, however, does go into hypoglycaemic coma and convulsions when it is fasted (Sampson, Taylor & Smith, 1955).

3

METABOLIC RATE, BODY TEMPERATURE, AND THERMAL INSULATION: CALORIMETRY

THE relation between an organism and its environment is influenced probably to a greater degree by temperature than by any other factor. In the case of the poikilotherm, the environmental temperature determines the temperature and therefore the level of activity of the organism. The homeotherm, on the other hand, has developed the ability to maintain, in the face of environmental fluctuation, a relatively stable internal temperature within a range largely determined (a) by the animal's maximum capacity to produce heat, (b) by its maximum thermal insulation, and (c) by its ability to dissipate heat. Even so, in addition to inducing thermoregulatory responses, the temperature of the surroundings plays a large part in the homeotherm's complex behaviour patterns in the search for food, shelter, territory, and reproduction. Other patterns of behaviour, while not immediately related to environmental temperature, have arisen in consequence of thermoregulation and the independence of the environment which this confers on the organism. Man, in particular, has enormously extended the types of environment which he can explore by interposing protective devices between himself and the surroundings. He has thus provided himself with an immediate environment, a micro-climate, which enables him to operate far beyond the limits which would otherwise be imposed on him by his capacity for physiological adaptation.

The relation between the organism and those attributes of the environment which influence heat exchange is considered at some length in succeeding chapters. In physiological terms there are three interrelated quantities to be considered: metabolic rate (taken to have the same meaning as rate of heat production), and both the temperature gradient and thermal insulation between organism and environment. It may be useful at this stage to examine the basic form which this relation takes, and its general importance for the thermoregulatory mechanisms of the homeotherm (Barnett & Mount, 1967).

BODY-TEMPERATURE, METABOLISM, AND THE ENVIRONMENT

It is undoubtedly easier to get some sort of measure of temperature of an organism than of heat flow from it, which is perhaps one reason why records of temperature greatly exceed in number those of heat flow. For a temperature gradient to exist, a flow of heat must occur, but what often happens is that temperatures are quoted without reference to metabolism. This may limit the usefulness of the information in the same way as would be the case if, say, the blood concentration of a metabolite were quoted without any indication being given of its rate of utilization. An animal in the cold may have a rectal temperature similar to that of an animal in the warm, but the first animal may have a metabolic turnover several times that of the second. The precedence given in measurement and interpretation, whether to temperature or to heat flow, depends on the priority of interest in the particular problem on hand. Thermoregulation may be taken for granted, within the obvious limits, by those who are primarily concerned with the animal's energy exchanges; while, for those concerned with control systems, variations in temperature of the system observed are of first importance, although the magnitude of the heat flow involved may also be taken into account.

Since the mid-1920's, the Yugoslav school, first under Giaja (1938*a*, *b*), and more recently under Gelineo (1959), has made highly significant contributions to the understanding of the temperature-metabolism relation, particularly in the new-born. Following their lead, a number of other workers, notably Scholander, Hock, Walters & Irving (1950) and Hart (1957), investigated further the connection between environmental temperature and metabolic rate. The basic diagram of thermoregulation showing this connection is given in Fig. 3.1. It illustrates the rise in metabolic rate which takes place as the surroundings become cooler, between the limits set by the minimal and maximal rates of which the animal is capable. Cooling of the environment below the level corresponding to the maximum metabolic rate leads to a falling body temperature and death in hypothermia; warming above the zone of thermal neutrality, in which metabolic rate is at a minimum, leads to a rise in body temperature, and death in hyperthermia.

Between the two extremes, thermoregulation operates effectively to prevent a change in body temperature. Below the critical temperature, which is defined as the environmental temperature marking the lower end of the zone of thermal neutrality (that is, the zone of environmental temperature in which metabolic rate is at a minimum), thermoregulation is achieved primarily by an

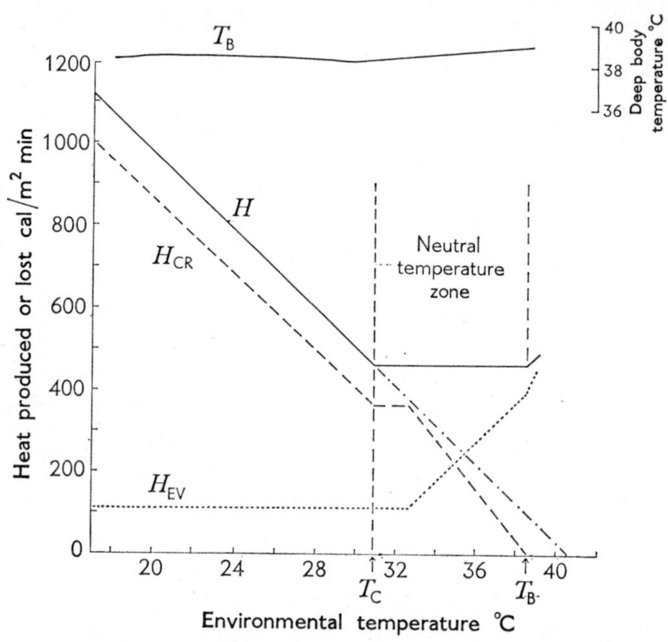

FIG. 3.1. Diagram showing heat production and heat loss related to environmental temperature in a homeothermic animal (kitten, three weeks old).

H : rate of heat production;
H_{CR} : heat lost by convection, radiation, and conduction;
H_{EV} : heat lost by evaporation;
T_C : critical temperature;
T_B : animal's deep body temperature

(from Hill, 1961, by permission of *British Medical Bulletin*).

increased metabolic rate, but depends too on other factors which increase the animal's thermal insulation, including peripheral vasoconstriction, pilo-erection, and postural changes. In the thermoneutral zone itself, thermoregulation is effected by vaso-motor control and sweating. The scale values to be assigned to the ordinate and abscissa of a diagram describing the thermoregulatory responses of any particular animal will vary according to species and according to its stage of development and state of acclimatization. For example, for the new-born pig the temperature scale might extend from 0 to 40°C, while for the older pig it might be from −40°C to 35°C under comparable standard conditions.

THERMOREGULATION

The essential feature of thermoregulation in the homeotherm is that the rates of heat production and heat loss are controlled in such a way that the body core temperature always remains relatively steady, when the environmental temperature changes, although the temperatures of the peripheral tissues and appendages, that is the body shell, may at the same time change considerably. The young pig, for example, has very little thermal insulation external to the skin surface, and consequently the environmental temperature may have marked effects on skin temperature. The temperature which is regulated is that of the core, not that of the surface. Indeed, if the surface temperature were to be maintained at a constant level, the animal would have to produce considerably more heat to keep the skin temperature the same while the surroundings became cooler. Another point to consider is the relation between deep body and surface temperature if skin temperature were maintained during ambient temperature changes. If the increase in heat production necessary to achieve this were to occur in the deep tissues, the deep body temperature would rise on account of the effective resistance to heat flow between the core and the skin due to the limitations on heat transfer by the blood flow to the skin, even with maximal peripheral vasodilatation. If, instead, the increase in heat production took place at the surface, this would not necessarily involve any change in the deep body temperature. The possibility of this happening physiologically is not very great, although the presence of metabolically active brown fat just below the skin in some animals (Smith, R. E., 1964; Hull, 1966) may have just this kind of significance. Experimentally, too, it has been possible to show that with external assistance in the form of an electrically heated wire garment (Wolff, 1958) fastened round the pig, it has been possible to keep both skin and rectal temperatures at their original levels as the environmental temperature falls.

At equilibrium, when the body temperature is stable, the rate of heat production equals the rate of heat loss to the environment. When the body temperature changes, heat is lost from or stored in the body, and a term to allow for the heat storage change must be included in any statement of the energy balance. This is sometimes an important part of the thermal balance. The camel is of special interest in this respect. It allows its body temperature to rise during the heat of the day, and to fall during the night; this appears to be an adaptation which allows the animal to conserve its body water. Some of the heat which would be dissipated by evaporation of water at the body surface of a more rigid homeotherm is stored, causing a rise in body temperature. This heat is then dissipated by non-evaporative means in the cooler part of the 24-hour cycle (Schmidt-Nielsen, 1964).

The regulation of core temperature poses the problem of how such control is achieved: by the central integration and processing of information from peripheral receptors, from centrally placed receptors, or from both peripheral and central. Careful examination of this highly complicated problem by many workers has revealed no single pattern of control. The hypothalamus is at the centre of the control system; this is largely agreed, although there is evidence that other parts of the brain have a thermoregulatory function (Bligh, 1966). The problem is complicated further by the suggestion that a raised rate of metabolism in the brain itself accompanies the raised rate of metabolism of the organism as a whole when exposed to cold (Donhoffer, Szegvari, Jarai & Farkas, 1959). If this is so, then the central temperature receptors in the hypothalamus which are considered to play a large part in thermoregulation may be influenced by purely local thermal changes in the control centre itself. This local heating could serve as a damping device, preventing overshoot, and it could be incorporated in the building of a model to describe the thermoregulatory process. However, it should be appreciated that the inclusion of such a stabilizing process so complicates the model that the virtue of simplicity, which the model otherwise has, is lost. One of the dangers of building increasingly complex physical or mathematical models is that they can be made to function with such authenticity that they appear to be the things they represent, and attention is then diverted from biological reality to the more amenable model.

The subject of thermoregulation has been reviewed extensively by Hardy (1961), Thauer (1961), and von Euler (1961), and most recently by Bligh (1966). Bligh points out that the evidence for the dependence of thermoregulation on hypothalamic control is now almost undisputed, although there are plenty of questions, particularly concerning the distribution and influence of thermo-receptors, which are still open and are the subject of considerable discussion. He considers that a coarse form of temperature control, with upper and lower limits which are not exceeded, may be common to all mammals, and that fine control may have developed separately in different groups of mammals. This could lead to a diversity of control systems as a result of many combinations between hypothalamic and extra-hypothalamic mechanisms.

THERMAL INSULATION

The insulation of an animal is most usefully considered in two ways. The first concerns the factors which affect the insulation per unit area of the animal's surface; the second concerns the overall thermal insulation of the animal which, for a given body-ambient temperature gradient, governs its total non-evaporative heat exchange with the environment (Mount, 1966*a*; Barnett & Mount, 1967).

Laws of heat flow

Heat flow is proportional to the temperature gradient and to the thermal conductance of the medium through which the heat is passing. This is a statement of Fourier's Law, which may be expressed (Kleiber, 1961):

$$H = \lambda \frac{A}{L}(T_1 - T_2)$$

where H = rate of heat flow,

A = surface area,

L = thickness of medium through which heat is passing,

λ = thermal conductivity of medium,

$T_1 - T_2$ = temperature gradient across the medium.

This can be re-written in a form corresponding to Ohm's Law for the flow of electricity:

$$H = A \frac{(T_1 - T_2)}{R}$$

where $R = \dfrac{L}{\lambda}$

$\qquad$ = resistance to heat flow per unit cross-section area.

In this instance, R can be defined as the 'specific insulation', a concept which can be applied to the thermal insulation between an animal's core and the environment.

Insulation per unit area

The operative factors include both the animal's *specific insulation* (discussed by Kleiber, 1961), depending on peripheral vasomotor control, subcutaneous fat layer, and hair coat, and the conditions controlling the micro-environment, particularly air movement, and affecting insulation of the boundary layer of air in a naked species or of the trapped air in the hair or wool of a coated species. The precise meaning of specific insulation is difficult to define. On p. 131 of his book, Kleiber defines the specific insulation of a katathermometer in terms of the difference between the mean temperature of the instrument and the effective environmental temperature, divided by the heat flow. The same calculation can be applied to an animal when evaporative heat loss at the surface is taken into account, since the thermal insulation between the animal's surface and the surroundings is related only to the non-evaporative part of the heat flow, which takes place through the channels of radiation, convection, and conduction. As conditions become warmer, an increasing part of the animal's heat loss is by way of the evaporative channel. Evaporative heat loss is not involved in the relation between thermal insulation and temperature difference since it is governed by different laws. The heat brought from the animal's deep tissues to its surface is all necessarily non-evaporative, however. Thus for the calculation of thermal insulation, the total heat transfer from the animal's core to the surface is used as the basis for tissue insulation, and only the non-evaporative fraction for the calculation of the thermal insulation external to the skin. Fig. 3.1 indicates the evaporative and non-evaporative components of heat loss in relation to the thermo-neutral zone, and the increased thermal conductance associated with peripheral vasodilatation is shown by the steeper slope of H_{CR} above about 33°C environmental temperature.

However, after defining the specific insulation of a dry kata-thermometer in terms of the temperature gradient between the instrument and the environment, Kleiber goes on (p. 143 of his book) to define the specific thermal insulation of an animal in terms of the difference between rectal and surface temperatures. What is the surface temperature of an animal with a coat? Manifestly not that of the skin: if it is the temperature of the outer surface of the coat, does this mean the outer edge of the main coat, or the outer edge of the associated boundary layer of air? This in turn means the air temperature, or, more strictly, the surrounding operative temperature (see Chapter 9).

To avoid this difficulty, and to take account of the change from the total heat flow between core and skin to only the non-evaporative flow from skin to environment, it would be preferable to refer to specific insulation either as *internal* (from core to skin) or *external* (from skin to environment). In the pig, which loses very little water from the skin surface even under hot conditions, the internal and external heat flows per unit area are similar. In

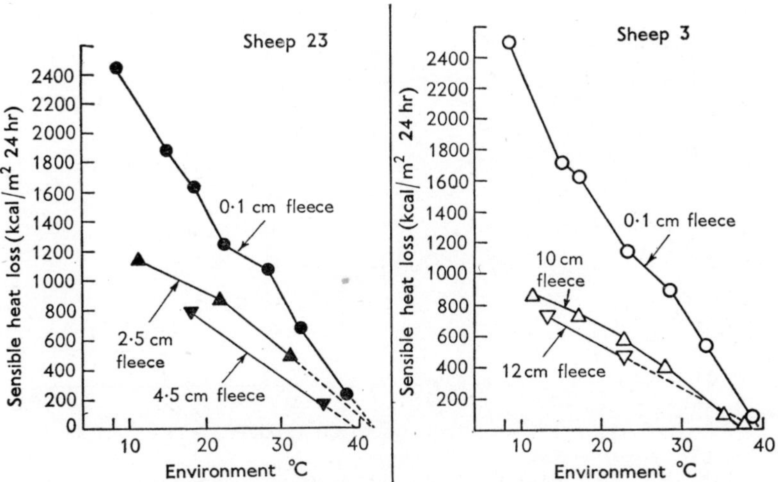

FIG. 3.2. The effects of length of fleece on non-evaporative heat loss from two sheep over a range of environmental temperature (from Blaxter, Graham & Wainman, 1959, by permission of *Journal of Agricultural Science*).

man, however, evaporative heat loss from the surface under hot conditions can amount to many times the resting metabolism, and in other animals the evaporative component can also be considerable. The introduction of a discontinuity at the actual skin surface in respect of treatment of heat flow and thermal insulation, regardless of coat or clothing, extends the usefulness of the concept of specific insulation. This means that factors in the environment, such as air movement, affect the external specific insulation.

In the animal with a coat of either hair or wool, this is the major factor in specific insulation, the thermal insulation offered by the coat depending on the air trapped between the fibres. This is illustrated in Fig. 3.2 which shows the effect of fleece length on heat loss in sheep. Blaxter & Wainman (1961) have calculated that a large part of the increase in insulation in steers below the critical temperature is due to pilo-erection. Tregear (1965) has made

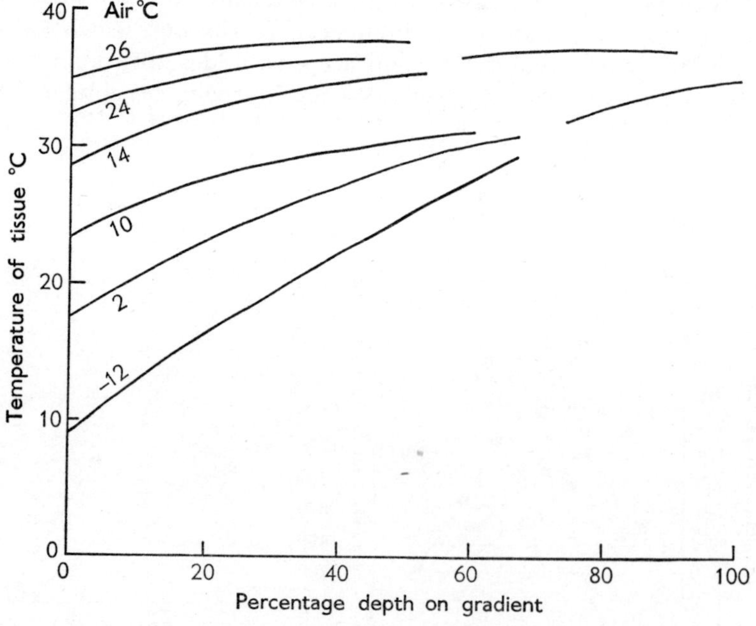

FIG. 3.3. Temperature gradients through superficial layers of pigs in air at several temperatures from February to August. Owing to variation in the actual thickness of the tissue gradient in different seasons, the abscissa represents the percentage of depth of the temperature gradient (from Irving, 1956*a*, by permission of *Journal of Applied Physiology*).

measurements of the thermal insulation of various skins, including that of the pig.

When a pig with a considerable layer of subcutaneous fat is exposed to cold, peripheral vasoconstriction is very effective in producing a high specific insulation. This was clearly shown by Irving's (1956*a*) pigs in Alaska; tissue temperatures were measured at various depths below the skin surface, using thermocouples sealed in hypodermic needles (see Fig. 3.3). The depth of the subcutaneous temperature gradient, that is from the skin surface to the body core, increased as the pigs became fatter in winter. The superficial tissues provide insulation which decreases in an inward direction. When there is no fat layer, however, as in the new-born pig, although peripheral vasoconstriction occurs in the cold the increase in insulation is small (Mount, 1964*a*), because although the cooled outer layer of tissue acts as an insulator it is not so effective as fat, partly because fat has a lower thermal conductivity than other tissues (Burton & Edholm, 1955). Fat is a functional insulator in a naked species like the pig (Ingram, 1964*b*) since not only does it work well in the cold, it also offers little thermal insulation in hot conditions because, as was shown for man by Miller & Blyth (1958), it can then be bypassed by blood flowing from deeper to surface tissues through greatly dilated blood-vessels.

Overall insulation

The overall impedance to heat flow from organism to environment is not necessarily predictable from knowledge of the specific insulation. Posture modifies heat flow, depending on whether the attitude is flexed or extended, an effect demonstrated convincingly by Benzinger & Kitzinger (1963) in a rapidly responding gradient layer calorimeter (see Fig. 3.4). When an animal becomes active it exposes a large effective heat exchange area to the environment and increases air movement over its surface; the effect of the latter is accentuated by convective heat loss taking place more readily from small parts, such as limbs, than from the trunk (Hardy, 1949). These factors cause a rise in metabolic rate in addition to the increase resulting from muscular activity.

Counter-current heat exchange

Counter-current heat exchange between arteries and veins, due to vascular arrangements in the limbs, and particularly in some aquatic mammals (Scholander, 1958), affects the overall heat loss by transfer of heat to venous from arterial blood before the arterial blood reaches the surface of the animal. This is important in animals in a cold environment, since although many species adapted to such an environment have thick fur coats, their legs

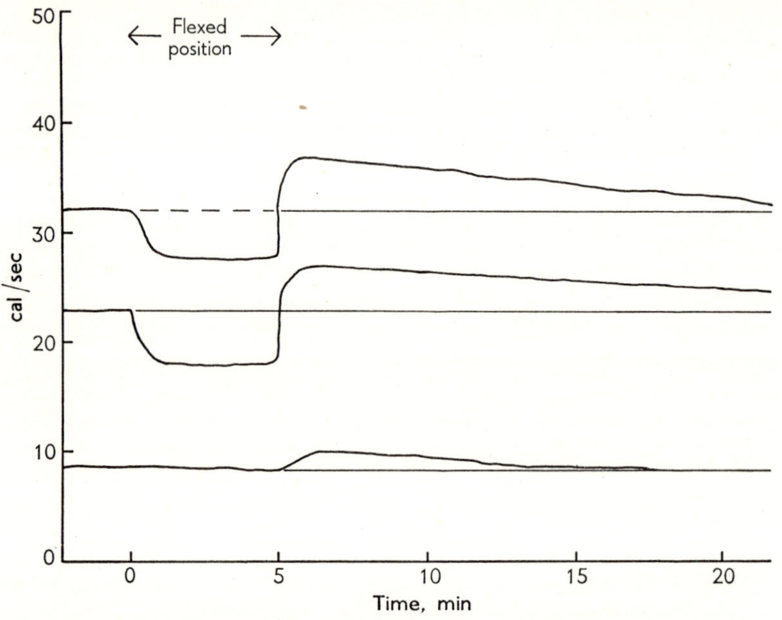

FIG. 3.4. The influence of posture on the rate of heat loss from the human subject. From above downwards, the traces represent the total calorimeter output (including the ventilatory circuit), the calorimeter output, and the output of the ventilatory circuit, from a gradient layer calorimeter. At time O, the subject adopted a flexed position, followed by re-extension at time 5 minutes. Heat loss is decreased markedly by flexion; there is an increased output by the ventilatory circuit following re-extension, due to the evaporation of water accumulated on the skin surface during flexion (from Benzinger, Huebscher, Minard & Kitzinger, 1958, by permission of *Journal of Applied Physiology*).

and appendages are poorly insulated, and if these were kept warm the heat loss would be very great. At low ambient temperatures, down to $-10°C$, less than 10% of metabolic heat is lost from the legs of herons and gulls, whereas at $+35°C$ almost the entire heat production is dissipated through the legs. The naked legs of these birds are very important heat exchangers. In water they lose four times as much heat as in air at the same temperature (Steen & Steen, 1965*a*). Another example of a thermoregulatory appendage occurs in the beaver. When totally surrounded by air, the beaver can maintain a normal rectal temperature of 37°C if the ambient temperature does not rise above 20°C. In warmer conditions the animal becomes hyperthermic. When the beaver is allowed to keep its tail in cool water, however, with its body in air, it maintains a normal rectal temperature at an ambient temperature of 25°C, losing 20% of its total heat production through its tail (Steen & Steen, 1965*b*).

In man, the temperature in the brachial artery can fall by as much as 0·3°C/cm as a result of heat transfer to venous blood (Bazett, Love, Newton, Eisenberg, Day & Forster, 1948), and a corresponding heat exchange may occur in the legs of pigs and

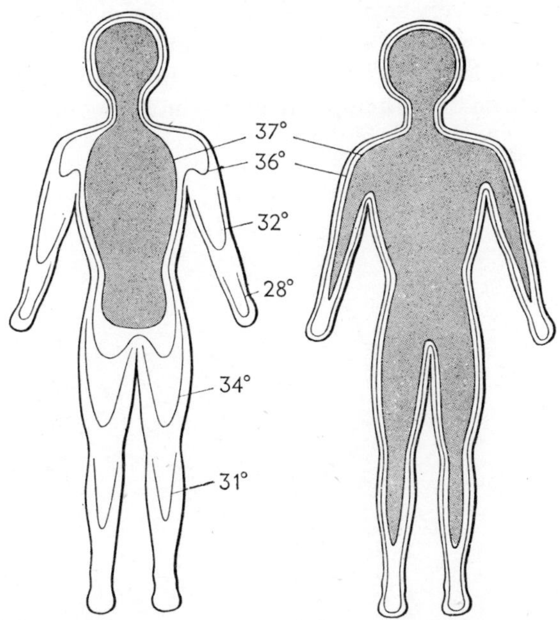

FIG. 3.5. Diagrammatic isotherms in the human body exposed to cold (left) and warm (right) conditions. The core of deep body temperature (stippled) shrinks in the cold, leaving a peripheral shell of cooler tissue (from Aschoff & Wever: 'Kern und Schale im Wärmehaushalt des Menschen'. *Die Naturwissenschaften*, Jg. 45, S. 477–485. Berlin-Göttingen-Heidelberg: Springer 1958).

other animals. Counter-current heat exchange no doubt contributes to the maintenance of a cooler shell of tissue around the warmer body core, in this way reducing heat loss to a cold environment and so effectively increasing the total thermal insulation (Fig. 3.5).

Body cooling rates

Kleiber (1961) has pointed out that the cooling of a body in air is approximately proportional to $W^{-1/3}$ (W = body weight). He arrives at this by assuming first that the rate of heat loss is proportional to $W^{2/3}$, which is itself approximately proportional to surface area. If the heat capacity is proportional to body weight, then the rate of cooling of a given body is proportional to $W^{2/3}/W = W^{-1/3}$; as the body size increases, so the rate of cooling decreases, but at a progressively diminishing rate. Thus the heat input required per unit body weight for the maintenance of body temperature falls due to an increase in body size alone.

Conversely, at very small body sizes, an animal's rate of heat loss per unit of body weight becomes very large if its body temperature is to be maintained by its own efforts. The tiny young of the mouse, for example, do not attempt to maintain homeothermy outside

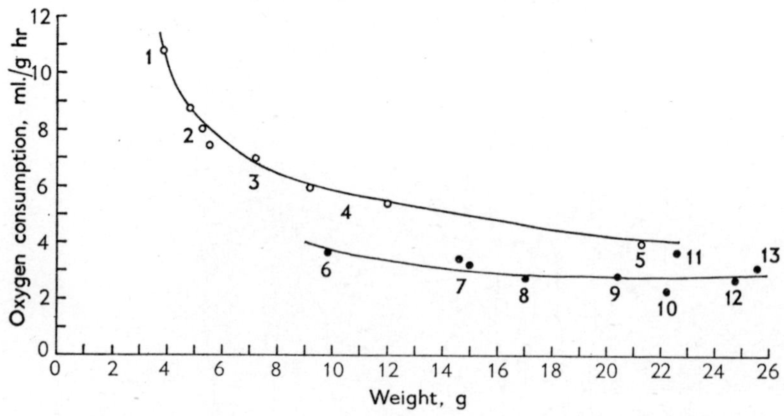

FIG. 3.6. The oxygen consumption rates of shrews (open circles) and mice (solid circles). 1, masked shrew; 2, wandering shrew; 3, Monterey shrew; 4, Sonoma shrew; 5, short-tailed shrew; 6, harvest mouse; 7, Kangaroo mouse; 8, house mouse; 9, Gapper's vole; 10, deer mouse; 11, Pine mouse; 12, white mouse; 13, Rhoads vole (from O. P. Pearson, by permission of *Science*, Vol. 108, p. 44, Fig. 1, 9 July 1948).

very narrow limits: their body temperatures tend towards the temperature of the surroundings. The humming-bird lapses into a state of cool torpor at night (Lasiewski, 1963), and the correspondingly lower metabolic rate makes a much smaller demand on its energy reserves. Pearson (1948) has utilized the shrews to investigate the lower limits of mammalian size compatible with homeothermy, and finds that 2·5 g is about the smallest size which can be expected for a homeotherm. At that body weight, the rate of oxygen consumption becomes very large (Fig. 3.6) and adequate food gathering for energy purposes is impossible. Shrews apparently have higher metabolic rates than mice of the same size.

CALORIMETRY

The first useful measurements of animal heat were made in the second half of the eighteenth century, in conjunction with the development of calorimetry in the study of chemical combustion. The early development of animal calorimetry is interesting in illustrating the relation between theory and experiment in the evolution of a new approach. This can be best appreciated by comparing the early work of Crawford in Scotland with that of Lavoisier in France. The first measurements in animal calorimetry appear to have been carried out in Glasgow in the summer of 1777 by Adair Crawford (1779, 1788). He described his work at a meeting of the Royal Medical Society in Edinburgh in the following winter.

He was a pupil of Joseph Black, and he made his calorimeter at Priestley's suggestion. The calorimeter depended in principle on the measurement of the rise in temperature of a jacket of water surrounding the reaction chamber, or animal chamber. Since heat loss from the calorimeter was largely prevented by insulation, the rise resulted from heat flow from the chamber. Crawford demonstrated that:

the quantity of heat produced when a given quantity of pure air is altered by the respiration of an animal is nearly equal to that which is produced when the same quantity of air is altered by the combustion of wax or charcoal, and that when an animal is placed in a cold medium, it phlogisticates a greater quantity of air in a given time than when it is placed in a warm medium.

At this time the phlogiston theory was still accepted, and the theoretical treatment of experimental results suffered accordingly. It was a great theoretical advance, therefore, when Lavoisier & Laplace (1780) translated combustion into terms of oxygen consumption, and slightly later, with their ice calorimeter, established a sound basis for animal calorimetry in relation to chemical combustion. The ice calorimeter depended on the collection of water from melting ice in a jacket surrounding the animal chamber. Haldane (1889) compared Lavoisier's calorimeter unfavourably with Crawford's apparatus on two counts: first, the error involved in collecting water from melting ice was greater than that in measuring a rise in temperature in a water jacket; and, second, the animal was necessarily exposed to a cold environment. Haldane commented:

> It would thus appear that Crawford's method was in his own hands a very exact one. It seems exceedingly doubtful whether any observer up to the present (1889) has obtained more accurate results in animal calorimetry.

However, whereas Crawford lived on with the phlogiston theory and its attendant obstructions to progress, Lavoisier advanced to 'oxygène' and the liberation of heat by chemical reaction. Crawford had come to the conclusion that the liberation of heat was due to changes in heat capacity of the reacting substances, and wrote (Crawford, 1788, pp. 370–1): 'We may conclude, therefore, that the sensible heat, which is excited in combustion, depends upon the separation of absolute heat from the air.' This publication was subsequent to that of Lavoisier and Laplace; Crawford discusses their concept of animal heat being due to 'chemical decomposition'. He disagrees with their view, and goes on to say (p. 380 of his book): 'Hence we may conclude, in general, that the evolution of heat from the air, in consequence of an alteration in its capacity, is the true cause of animal heat, and of that which is produced by the inflammation of combustible bodies.'

Although Crawford failed to understand the true significance of what was taking place, he demonstrated the increase in heat production which takes place in a guinea-pig in the cold. In spite of this achievement, however, he was unable to take the necessary step for the advancement of his subject, because he could think only in terms of entrenched theory. It was left to Lavoisier to establish the basis for future metabolic work. Thus although Crawford's observations were the more accurate, the development of metabolic studies in fact followed Lavoisier's recognition of oxygen and the realization of the similarity between chemical combustion and animal metabolism.

These early measurements depended on direct calorimetry, that is the measurement of heat loss. Two main types of direct calorimeters have emerged. The one depends on absorbing the animal's heat loss and measuring it as a rise in temperature in the absorbing medium, as was the case with Crawford's calorimeter; the other depends on the measurement of the temperature difference produced across a layer surrounding the animal as the result of heat flow from the animal to its surroundings.

The most celebrated instrument of the first kind was the Atwater–Rosa–Benedict apparatus, in which heat was removed in circulating water (Atwater & Benedict, 1905; and see Lusk, 1928). Similar methods were used by Armsby & Fries (1903) for cattle, and by Capstick (1921), Capstick & Wood (1922), and Deighton (1937) for pigs; the pig calorimeter was developed initially by Hill & Hill (1914). Also belonging to this class is the 'air' calorimeter of Kelly, Bond & Heitman (1963), which is similar to that of Auguet & Lefèvre (1929) in which the animals' heat loss is measured from the temperature rise in the ventilating air stream.

Recently, a large 'heat sink' calorimeter of the circulating water variety has been made at Babraham, and used for the continuous measurement of heat loss from groups of pigs living in the apparatus for several weeks at a time (Mount, Holmes, Start & Legge, 1967). The calorimeter is mounted in a shell-space maintained at the same temperature as the calorimeter interior, which is equipped like a pig pen (Fig. 3.7). The pen temperature is held at a given level by an automatically operating water-circulated heat exchanger controlled by the temperature of the exhaust air from the pen. The non-evaporative heat loss from the pen is given by the inlet-outlet water temperature difference across the heat exchanger. This temperature difference is continuously recorded, and is compared with the temperature difference produced by a standard heat input at a point downstream from the heat exchanger (Fig. 3.8), so allowing heat loss from the pen to be calculated by proportion. This removes the need to measure the water flow and calculate the heat transfer; the continuous recording of both temperature differences, from the heat exchanger and from the standard heat input, makes it unnecessary to have a highly stabilized water flow, and variations in the specific heat of the

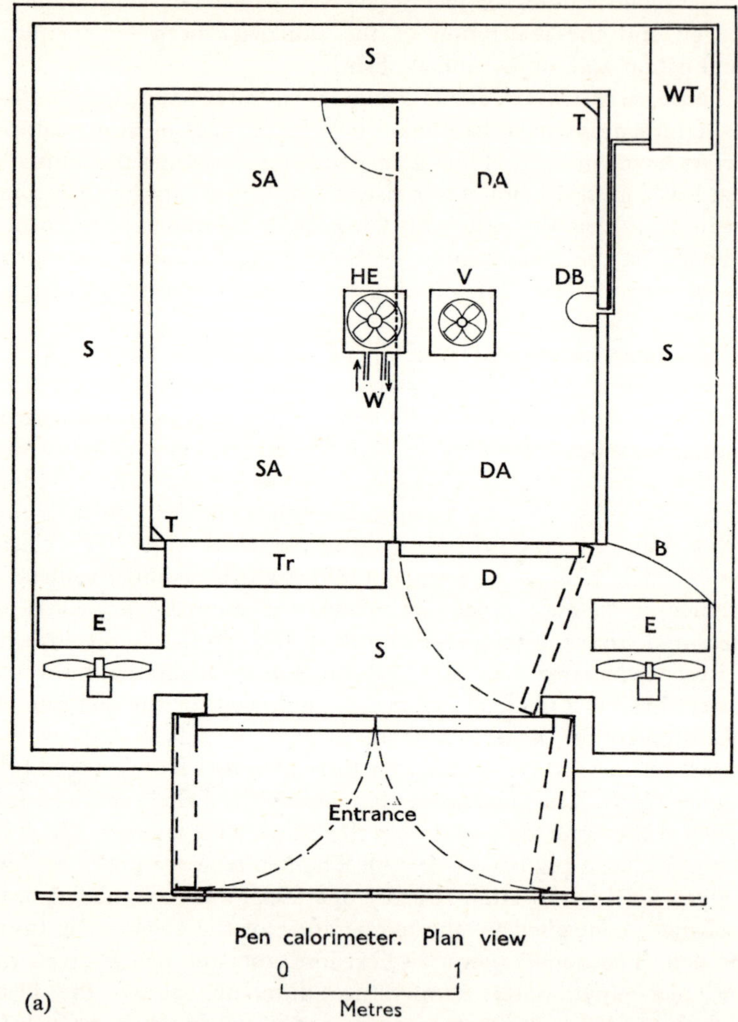

FIG. 3.7. (*a*) Horizontal and (*b*) vertical projections of the Babraham pen calorimeter. S, shell-space; E, refrigerator evaporator and fan; WT, drinking water tank; B, baffle; D, dunging passage door; Tr, feeding trough; T, thermocouple; SA, sleeping area; DA dunging area; DB, water bowl; HE, heat exchanger; W, circulating water; V, ventilation exhaust fan (ventilation rates: 50,000–83,000 l./hr); P, pigs; F, slatted floor; Dr, drain; C, ceiling of pen; A, air; WB, wet-bulb thermocouple assembly; RT, resistance thermometer (from Mount, Holmes, Start & Legge, 1967, by permission of *Journal of Agricultural Science*).

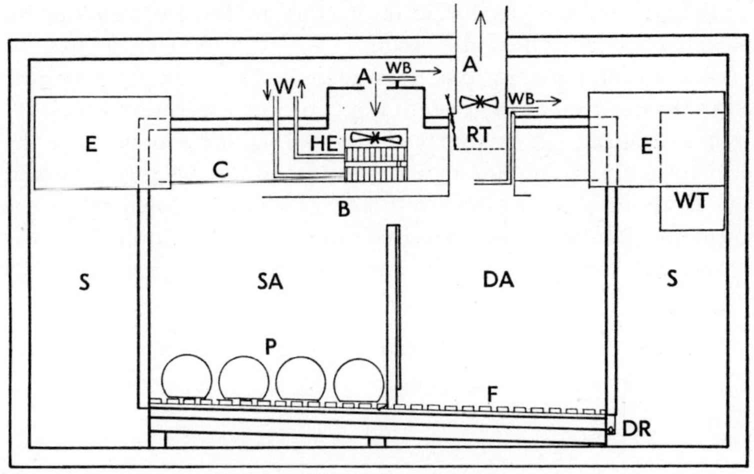

Pen calorimeter. Front elevation

0   1

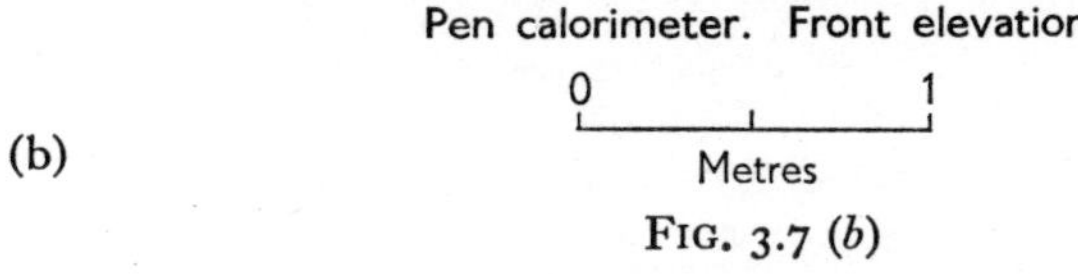

Metres

(b)

Fig. 3.7 (*b*)

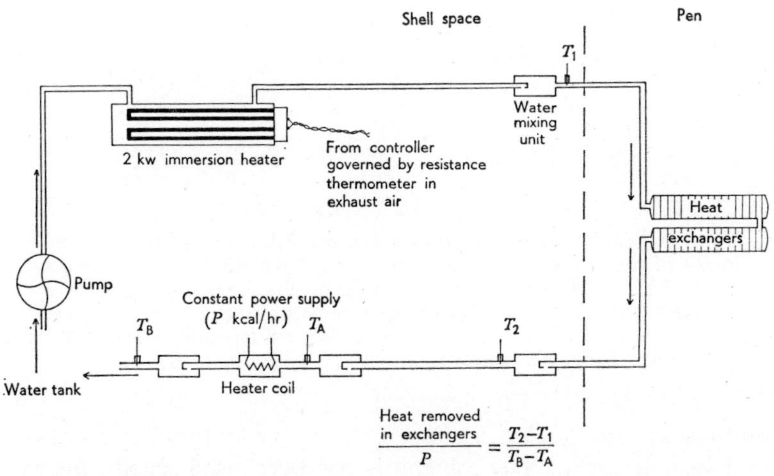

$$\frac{\text{Heat removed in exchangers}}{P} = \frac{T_2 - T_1}{T_B - T_A}$$

Fig. 3.8. Diagram of water-circulating system in the Babraham pen calorimeter. Water mixing units are placed in the water line immediately before the points at which temperatures T_1, T_2, T_A, and T_B are measured by thermocouples (from Mount *et al.*, 1967, by permission of *Journal of Agricultural Science*).

circulating medium, as for example due to the introduction of antifreeze, do not affect the result.

A calibration curve for the measurement of non-evaporative heat loss in the calorimeter is given in Fig. 3.9. Any correction necessary for heat transfer across the walls of the calorimeter and in the ventilating air rarely amounts to more than 2% of the total heat loss. Evaporative heat loss from the pen is calculated from the ventilation air-flow rate and the water vapour contents of the ingoing and outgoing air. The mean coefficient of variation from

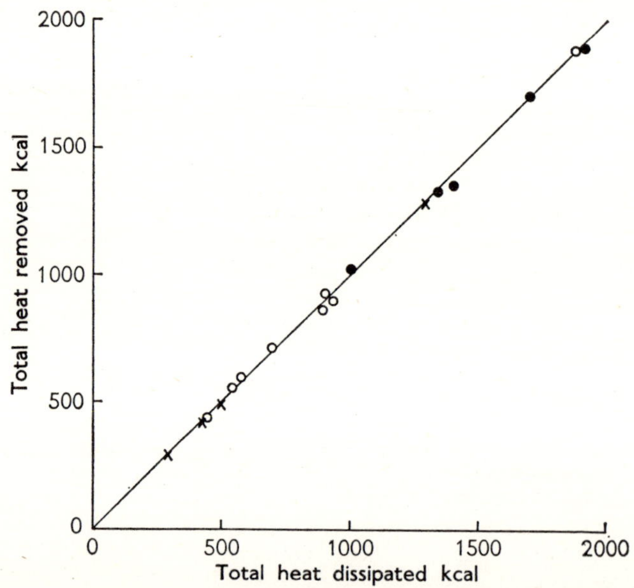

FIG. 3.9. Calibration of the Babraham pen calorimeter for measurement of non-evaporative (sensible) heat dissipation. The test runs ranged in length from 1·3 to 6·0 hr, and were carried out at 9 (○), 20 (●) and 30°C (×). The linear regression equation is: Heat removed (kcal) = 0·99 (heat dissipated) + 5·8 (from Mount *et al.*, 1967, by permission of *Journal of Agricultural Science*).

calibration tests, including both evaporative and non-evaporative heat loss, is 2·3%. The apparatus has been used chiefly for the measurement of 24-hourly heat loss from groups of pigs, but its rate of response is high enough to allow short-term changes to be followed, as is clear from the response to a step change in heat input (Fig. 3.10).

The other method of direct calorimetry was first used by Richet (1889) and Rubner (1894), who estimated the rate of heat flow from the temperature difference in concentrically arranged air spaces around the calorimeter. Day & Hardy (1942) used the same principle for the measurement of heat loss from babies, and Prouty, Barrett & Hardy (1949) made an animal calorimeter on similar lines. Their apparatus consisted of inner and outer copper cylinders, fixed in relation to each other. Heat flow from inside the

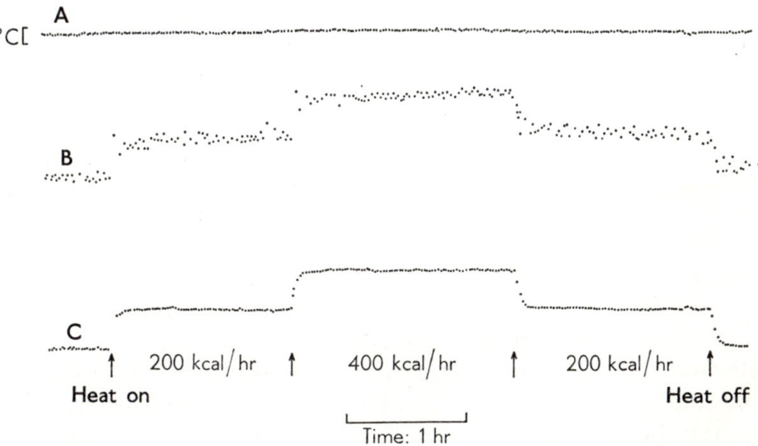

Fig. 3.10. The response of the Babraham pen calorimeter to changes in the rate of heat dissipation. A: calorimeter air temperature; B: heat removal rate by heat exchanger; C: rate of heat dissipation in calorimeter, increased in steps from 0 to 200 to 400 kcal/hr (from Mount *et al.*, 1967, by permission of *Journal of Agricultural Science*).

inner cylinder produced a temperature difference across the layer of air between the two cylinders, so that in practice the heat flow was determined as directly proportional to this temperature difference, which was measured by thermocouples. An electric fan blowing air on the outer cylinder maintained the outer shell at room temperature, and in this way the shell and fan constituted a heat absorber of large capacity.

The concentric shell type of apparatus was succeeded by the accurate gradient layer calorimeter of Benzinger & Kitzinger (1949); Pullar (1958) has also described a gradient layer calorimeter. Hammel & Hardy (1963) have used a gradient layer calorimeter for a dog, coupled with oxygen and carbon dioxide measurement, for the assessment of thermoregulatory responses. In these calorimeters the temperature difference across a thin layer of material lining the inside of the chamber is measured by numerous thermocouples distributed over its whole area, so that the animal's heat loss is accurately integrated.

Distinct from these direct calorimetry approaches, which may be termed heat sink and thermal gradient methods, are the systems of indirect calorimetry, in which the material exchanges accompanying metabolism are used to compute an animal's heat production. These again fall into two main groups: one is respiration calorimetry; the second is dependent on the carbon and nitrogen analysis of food intake and may often be combined either with respiration calorimetry or with carcass analysis (Blaxter, 1965).

Respiration calorimetry in the form of measurement of oxygen consumption is very widely used to determine heat production either in closed-circuit or open-circuit systems. Developed from the early apparatus of Reignault & Reiset (1849), the closed-circuit method, in which the animal is totally enclosed in a chamber forming part of the gas circuit has led to the development and use of apparatus of high accuracy (Brody, 1945; Alexander, 1961*a*; Blaxter, 1962*a*). Open-circuit systems have evolved from designs originally used by Haldane and by Pettenkofer (see Kleiber, 1961, p. 63); the use of the diaferometer and other devices for measuring concentrations of oxygen and carbon dioxide is now making the open-circuit method both easier and more accurate in use.

The various methods of calorimetry used in the past, and those in current use, have been described adequately in many text-books and other publications; reference may be made to Lusk (1928), Brody (1945), Kleiber (1961), and Blaxter (1962*a*). Attention is drawn to them here so as to indicate the range of approach to the measurement of heat production and heat loss in the whole animal, because this is obviously of the greatest importance in assessing the effects of environment on energy exchange. The choice of the method to be employed in any particular case depends on the sort of investigation in hand; in the Environmental Physiology Laboratory at Babraham all the modes of calorimetry listed above have been used, either singly or in combination, during the last few years. In other laboratories, the methods used for studies on pigs include: gradient layer calorimetry on young animals (Cairnie & Pullar, 1959), indirect calorimetry using face mask and spirometer (see Fig. 12.5, p. 318, in Brody,

1945), the Haldane open-chain principle (Zausch, 1965), and for pigs up to 90 kg body weight an open circuit respiration apparatus (Thorbek & Neergaard, 1965).

CONTROLLED ENVIRONMENT

Interest both in experiments on animals and plants under controlled environmental conditions and in the provision of facilities for this kind of work has increased throughout the world. The space designed for such a purpose may be termed a biotron, controlled environment room, climatic chamber, or simply a hot or cold room, and may vary widely in terms of the precision of environmental control, the number of factors controlled, and the range over which they operate. Information on the location and operational characteristics of controlled rooms used for research on man and animals is given by Henschel (1964) on the basis of replies to an inquiry to 100 scientists throughout the world known to be engaged in work on environmental physiology. Some of the climatic laboratories for farm animals have been described in detail (Findlay, McLean & Bennet, 1959; Heitman, Kelly & Hughes, 1949).

At Babraham a climatic laboratory has been built mainly for study of the environmental physiology of the pig. The layout of the building is shown in Fig. 3.11; since its facilities have proved suitable for a variety of work, they are briefly described. Controlled environment facilities are provided chiefly in the large laboratory on the left of the diagram. The climatic chambers (A) are made from panels of expanded polystyrene, 10 cm thick, faced with aluminium sheet or marine grade plywood, each panel of overall size 2·1 m × 0·92 m, easily carried by one man. These panels are bolted together with wood strips, and attached to overhead steel beams (B). The ceilings of the chambers are made of similar panels. The floor sections are built from 2·5 cm thick blockboard mounted on wooden beams, resting on the tiled laboratory floor, with the space between the beams packed with expanded poly-styrene. The wall sections are bolted to the floor sections. Doors to the chambers are made of similar panels, with light spring-catches fitted.

The area enclosed is divided under each laboratory beam, and down the centre (Fig. 3.11), by panels of expanded polystyrene 15 cm thick and faced on both sides by aluminium sheet. The components are so arranged that any section may be taken down completely without interfering with the rest of the structure.

Standard commercial type refrigerators fitted with automatic defrosting equipment are used for low temperature control, with precision thermostats; this combination is very effective. Most of

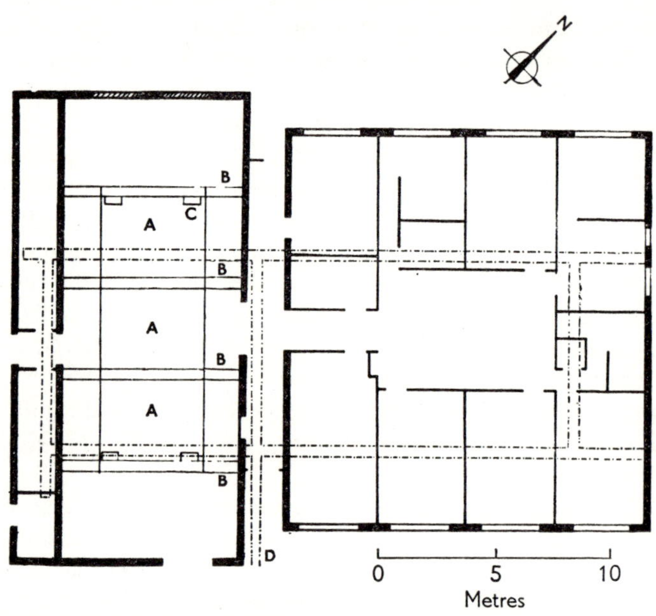

FIG. 3.11. Plan of the Environmental Physiology Laboratory at Babraham. A: climatic chamber; B: overhead steel beam; C: refrigerator evaporator; D: under-floor service duct.

the compressor units are of one horse-power capacity, and are coupled with fan-assisted evaporators (C) mounted in the corner of each room. Each fan runs continuously, and refrigeration and heating (strip heaters in front of the fans) are separately thermostatically controlled. Ventilation of the rooms is achieved by extraction fans operating through exhaust ducts, and can be adjusted by louvres.

When the door of any of the controlled chambers set at 5°C is closed, its air temperature is found to vary not more than ±0·5°C, and usually within ±0·3°C. At 35°C the total variation is only 0·2°C. At temperatures just around the ambient, that is 18–25°C, both heating and refrigeration are required to maintain thermal stability within the range of ±0·3°C.

In order to determine the effect the temperature in one room had on that next to it, one room was cooled to $-8.6°C$, while the adjacent room was heated to $40.1°C$. The temperatures on the aluminium surfaces of the dividing wall were $-7.5°C$ and $38.9°C$, differences of 1.1 and $1.2°C$ respectively from the corresponding air temperatures. Under less extreme conditions, wall temperature is within $0.5°C$ of air temperature.

These rooms have been in continuous use since 1962. They have proved very satisfactory for all the purposes for which they were originally designed, and have also been highly adaptable in their application for a variety of other experimental purposes.

4

THERMAL RELATIONS BETWEEN THE NEW-BORN PIG AND ITS ENVIRONMENT

BODY TEMPERATURE

THE rectal temperature of the mature pig is in the region of 39°C. When the pig is born, it also has a rectal temperature close to 39°C. In the first minutes following birth, however, the animal's temperature falls abruptly, and then, usually within half an hour, begins to rise again. Both the extent of the initial drop, and the time taken for the subsequent recovery, depend on the environmental conditions. In a cold environment the initial fall is greater than if the animal is born in the warm, the onset of the subsequent rise in body temperature is delayed, and the return to the original level takes longer. In the new-born pig, the mature level of about 39°C deep body temperature is normally reached in one day under usual conditions; this may be extended up to a week in a more adverse situation.

Fig. 4.1 illustrates body temperature changes in the post-natal period. Fig. 4.1*d* suggests that if a pig has recently been suckled it can withstand a two-hour exposure to cold as early in life as 12 hr after birth; if it has not recently been suckled, its ability to withstand cold is not good even at 48 hr (Pomeroy, 1953).

A considerable range of variation of body temperatures is found in different mammals following birth. At one extreme are the small and relatively immature rat and mouse, which are so easily chilled that until recently they were considered to be poikilothermic at the time of birth. The new-born rat's thermoregulatory capacity is in any case exceeded at environmental temperatures below about 30°C (Barić, 1953; Taylor, 1960). Above these temperatures, if it is fed, it shows the homeothermic type of response to cooling of the environment in that its metabolic rate increases; if it is not fed, or if it is kept at lower temperatures, it behaves like a poikilotherm, because its ability to produce heat does not satisfy the environ-

60

mental thermal demand. At the other extreme of thermoregulatory capacity is the reindeer calf, *Rangifer tarandus*, which shows a high level of cold-resistance (Hart, Héroux, Cottle & Mills, 1961). The domestic calf also shows a remarkable degree of independence of

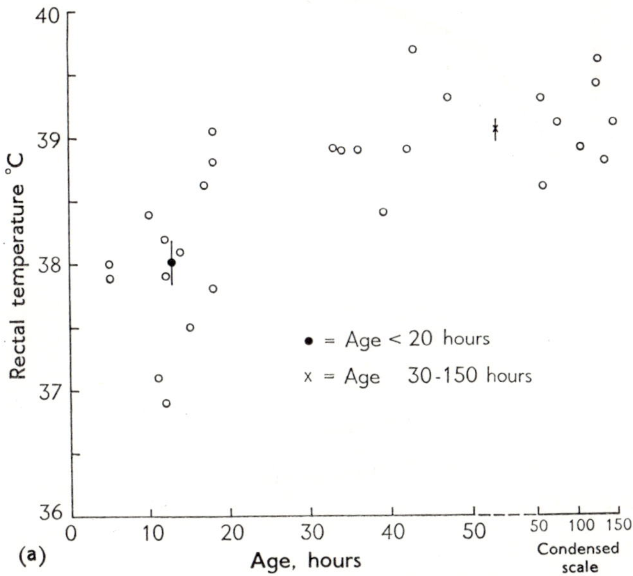

FIG. 4.1. Sets of observations of rectal temperature in new-born pigs during the period following birth. (*a*), individual observations and calculated first and second day means for Large White piglets (from Mount, 1959, by permission of *Journal of Physiology*). (*b*) and (*c*), rectal temperature in relation to age in warm and cold environments (from Newland, McMillen & Reineke, 1952, by permission of *Journal of Animal Science*). (*d*), the effects on rectal temperature of exposure of new-born pigs to about 3°C (×) compared with control pigs (⊙) left with the sow (from Pomeroy, 1953, by permission of *Journal of Agricultural Science*).

the environment (Roy, Huffman & Reineke, 1957), and the lamb has a body temperature close to the mature level within an hour or so of birth (Alexander & McCance, 1958).

METABOLIC RATE

The short-term measurement of oxygen consumption rate in the new-born pig has produced the results shown in Fig. 4.2. In

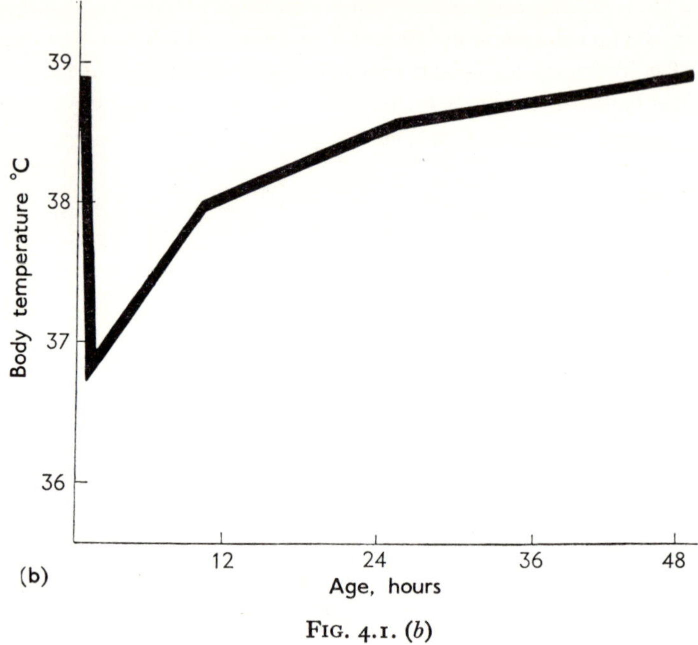

FIG. 4.1. (*b*)

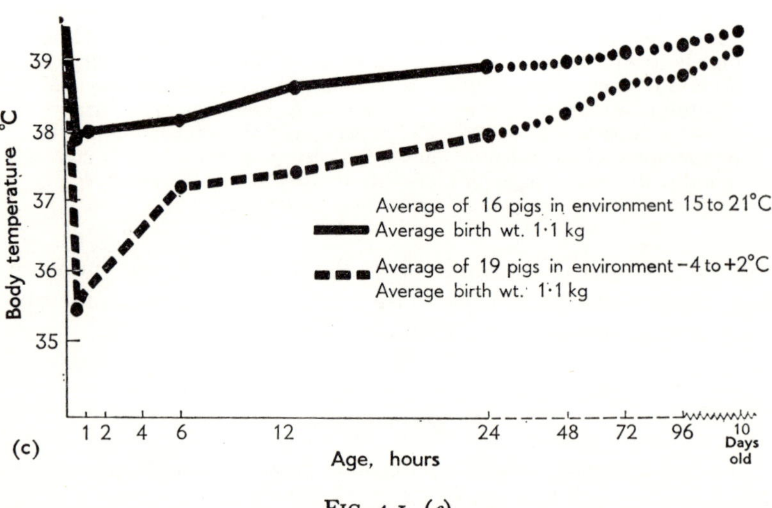

FIG. 4.1. (*c*)

order to make the measurements, the animal was placed by itself, free to move about and to adopt any posture, in a small metabolic chamber situated in a temperature-controlled space. The chamber contained soda asbestos, for the absorption of the carbon dioxide produced, and was connected to an accurate spirometer filled with

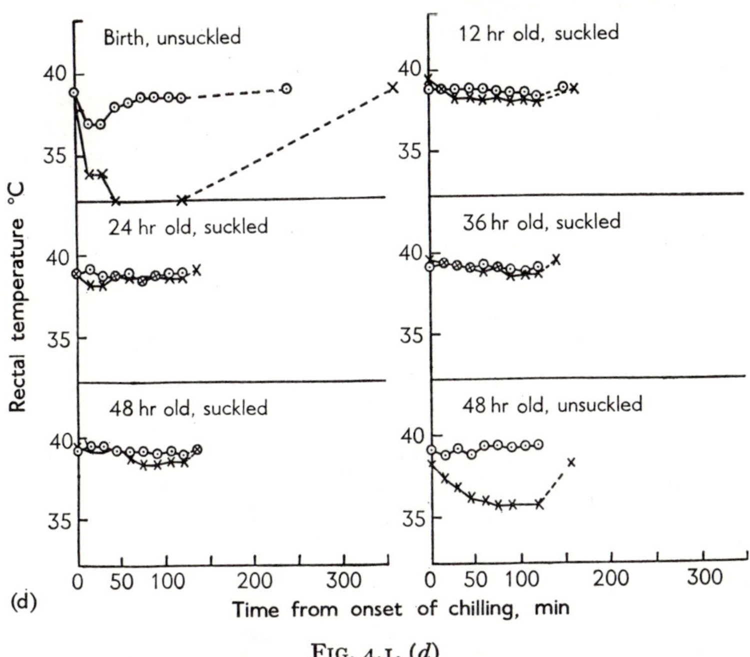

FIG. 4.1. (*d*)

oxygen. It was ensured that an oxygen concentration close to 21% was maintained in the air of the chamber. Following thermal equilibration at a given temperature, the rate of oxygen consumption was determined from the rate of descent of the spirometer over a period of half to one hour, with any necessary corrections applied for differences between the beginning and end of the run in respect of barometric pressure, temperature and humidity (measured by wet and dry bulb thermometers) in the chamber. Only very rarely did pigs treated in this way show any sign of disturbance; they rested quietly in the chamber throughout the period of observation.

The results in Fig. 4.2*a* show that pigs during the first day following birth exhibit the characteristically homeothermic type of response to cooling of the environment, in the form of a rise in metabolic rate; there is no significant change in mean rectal temperature at the environmental temperatures used in these experiments. The rise in metabolic rate takes place regardless of whether the animals have fed from the sow or not. The critical temperature lies in the region of 34°C. Exposure to 20°C results in a doubling of the thermoneutral level of metabolism, so that for the baby pig 20°C is a cold environment; Pullar (1963) obtained comparable results with new-born pigs.

After the first day the animals show a decidedly higher level of metabolic rate over the whole range of environmental temperature, although the critical temperature remains high (Fig. 4.2*b*). Com-

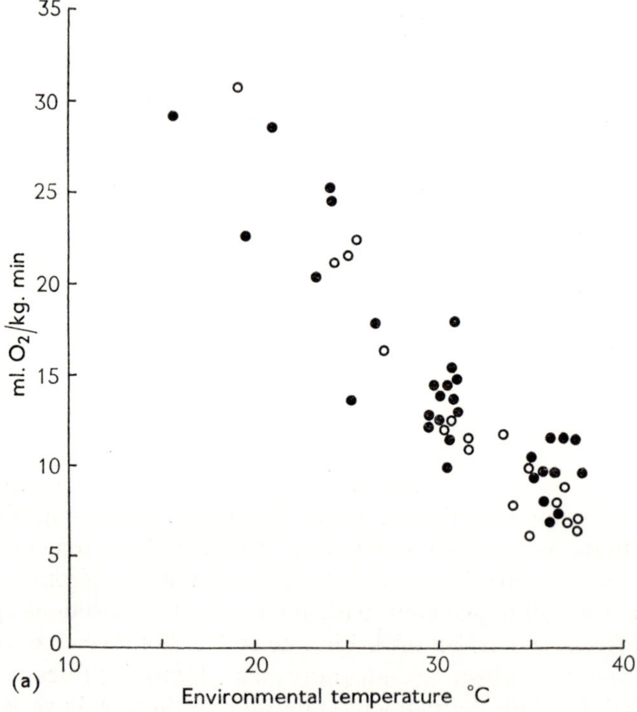

Fig. 4.2. The relation between environmental temperature and oxygen consumption in new-born Large White pigs.

(*a*) 10–18 hr of age. ○, not suckled; ●, suckled.
(*b*) ×, 1–2½ days old; +, 3–6 days old.
(*c*) mean values: the lower curve refers to pigs 10–18 hr old, and the upper curve to pigs 1–6 days old

(from Mount, 1959, by permission of *Journal of Physiology*).

parison of the first post-natal day with the rest of the first week
Fig. 4.2*c*) shows a significant difference in mean metabolic rate
over the temperature range used (Mount, 1959).

In common with other new-born mammals (Dawes, 1961), the

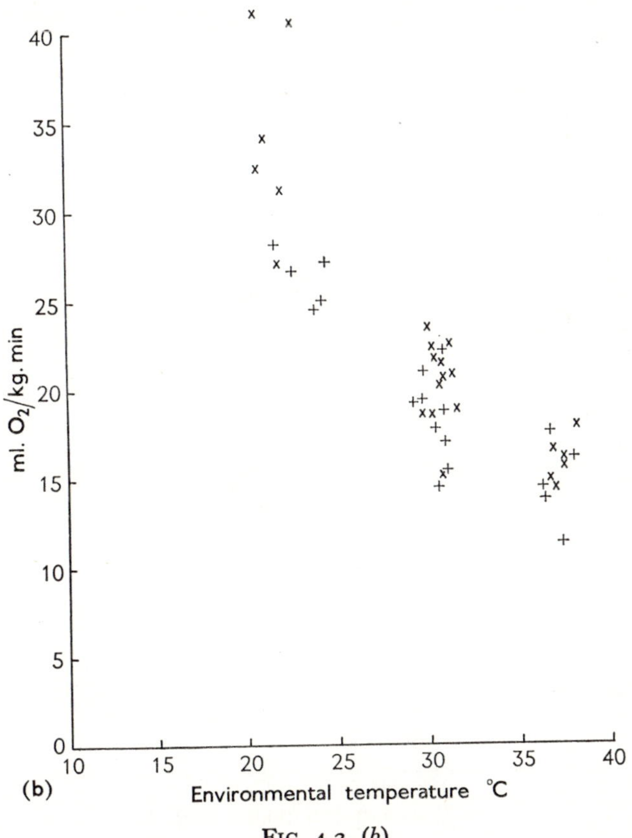

FIG. 4.2. (*b*)

new-born pig shows a fall in oxygen consumption when it is
exposed to a lowered concentration of oxygen in the inspired air. At
concentrations below 12%, the rate of oxygen consumption falls to
60–70% of the rate at 21% oxygen concentration; this effect has
been observed at environmental temperatures from 6 to 32°C
(Mount, unpublished observations).

Zhmurin (1959) has quoted some early work by Kudryavtsev
(1939, 1940), who found that metabolic rate in the piglet reached a

maximum 5–10 days after birth when expressed per kg body weight, and 10–20 days after birth when expressed per m² of body surface. Sokolova (1951), also referred to by Zhmurin, found the minimal metabolism of pigs in the first 10 days following birth to be 131 kcal/kg.24 hr.

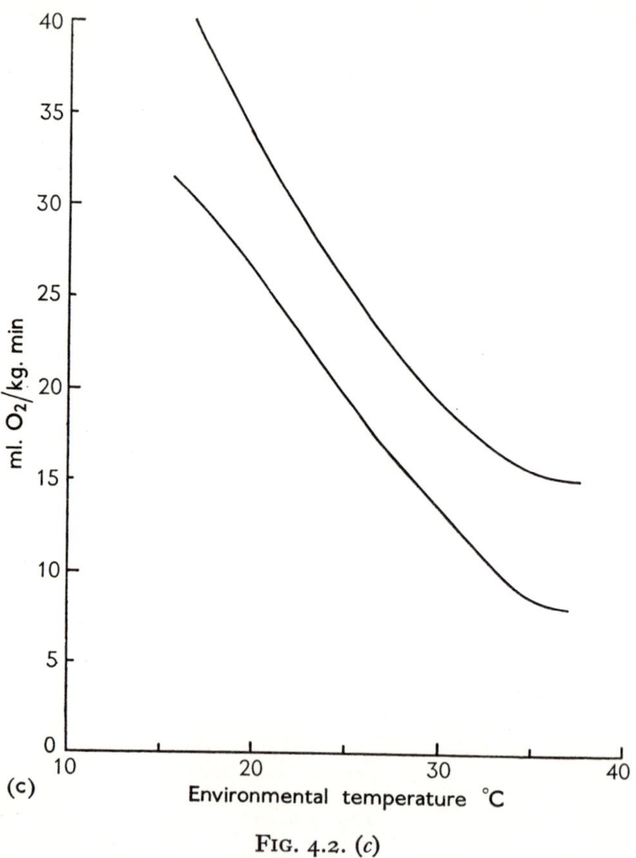

FIG. 4.2. (c)

BODY TEMPERATURE, METABOLIC RATE, AND THERMAL INSULATION

The question which arises is how the relation between heat production, body temperature and thermal insulation changes during the first few days following birth. The relation between these three quantities can be seen as analogous to Ohm's law, with the thermal insulation (resistance to heat flow) as the ratio of the body-ambient temperature gradient to the rate of heat loss. The heat flow referred to here is the non-evaporative part of the total heat flow. From the animal's core to the skin surface, the heat flow which is involved is the total flow, but from the skin to the environment part of the heat loss is by evaporation. The evaporative fraction is small in the pig at temperatures below the thermoneutral zone, being only about 10% or less of the total (see Chapter 6).

If heat loss is at a lower level in the immediately post-natal period than it is in the subsequent period, then either the body temperature is lower, or the thermal insulation is higher, or both.

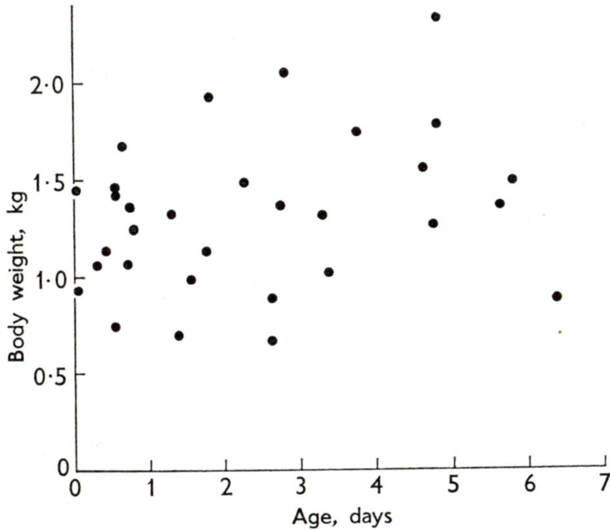

FIG. 4.3. The range of age and body weight of pigs used in the 5°C experiments.

The pig's lower metabolic rate immediately following birth may be due in part to the lower heat production demanded for the thermoregulatory processes associated with a lower body temperature. The mean rectal temperature rises from 38°C in the first post-natal day to 39°C on the second and subsequent days under the conditions which were used to obtain the results shown in Fig. 4.2c. If the increment in metabolic rate were due only to the 1°C rise in body temperature, the heat production increment would be about 15% at 30°C ambient temperature, and about 6% at 20°C. Instead, however, it is about 50% at 30°C and 30% at 20°C, so that it is decidedly greater than that required for the rise in body temperature alone. In order to seek an explanation for these changes, it is necessary to digress at this point to consider further experimental evidence.

METABOLIC RATE IN THE COLD

A further series of experiments at a lower environmental temperature, 5°C, helped to throw some light on the metabolic rate–body temperature relation. A number of new-born pigs were

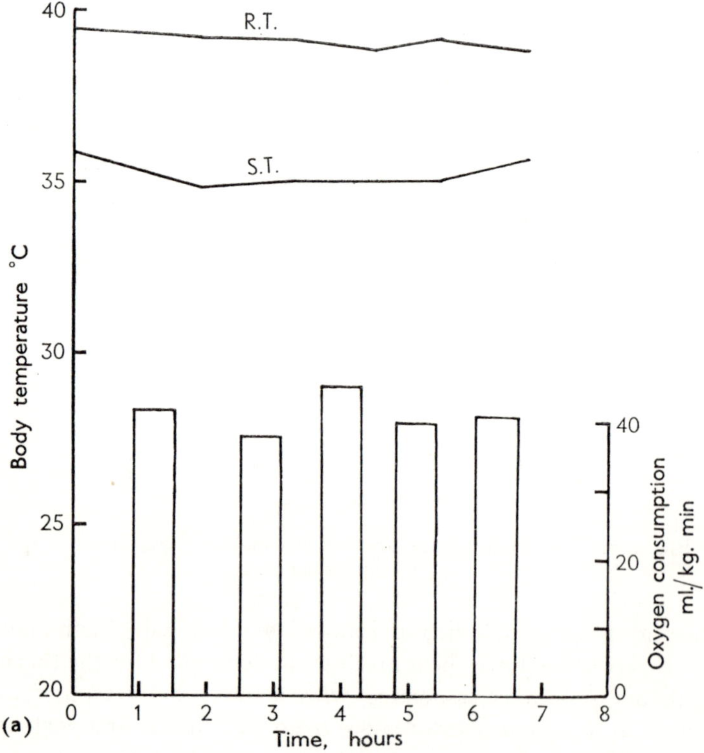

FIG. 4.4. Results from two experiments to show the variation between pigs in rates of change of rectal temperature, skin temperature and oxygen consumption, when the animals are exposed to 5°C environmental temperature. (a): pig aged 4 days 18 hr, weight 1·26 kg; (b): pig aged 1 hr, weight 0·93 kg. R.T.: rectal temperature; S.T.: skin temperature (from Mount, 1961, by permission of Ciba Foundation).

exposed singly, in a chamber at 5°C, under standard conditions, and oxygen consumption and rectal and skin temperatures were measured for continuous periods of up to seven hours in each case. The animals were selected so that they represented a wide range of body weight at each age (see Fig. 4.3). Some pigs maintained their temperature and heat production rate throughout the

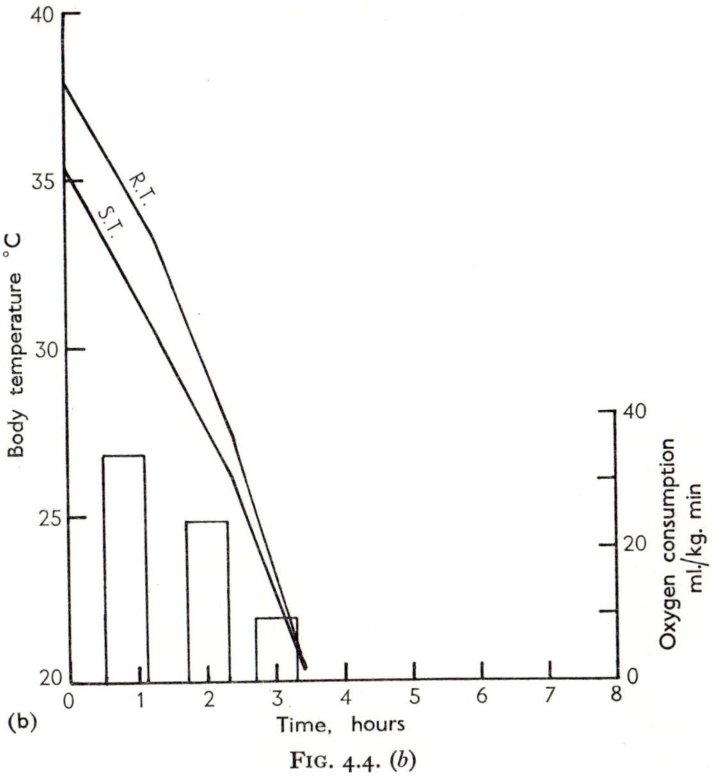

FIG. 4.4. (*b*)

period of exposure, whereas the temperatures of others fell markedly and they had to be rescued from hypothermia (Fig. 4.4). Fig. 4.5 shows the time-course of rectal temperature in the 31 pigs used in the series; on the whole, either a pig succeeded in producing enough heat in the face of the cold stimulus, or it failed rapidly; there were very few intermediate cases. In no case was the onset of hypoglycaemia a limiting factor in the animal's ability to withstand cold; indeed, blood sugar concentrations were usually above 100 mg/100 ml, perhaps due to adrenaline release stimulated by the cold environment. Silverman & Agate (1964) have noted similar tendencies to thermal 'successes' and 'failures' in new-born human infants, although of course under less extreme conditions.

THERMAL INSULATION AND THE CRITICAL TEMPERATURE

The results of these 5°C experiments varied considerably in respect both of between-animal variation in metabolic rate, and of

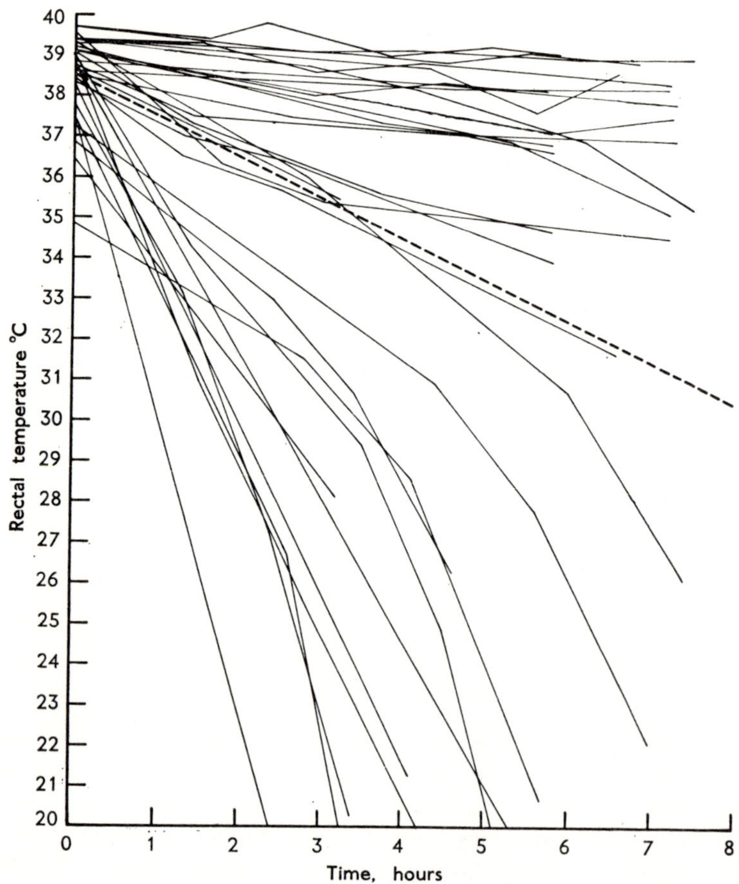

FIG. 4.5. Relations between rectal temperature and length of exposure to 5°C environmental temperature for 31 new-born pigs. The interrupted line corresponds to a rate of cooling of 1°C/hr (from Mount, 1963*a*, by permission of Federation Proceedings).

the course each animal followed during the experiment. This variation gave way to a surprising degree of uniformity when the results were plotted in the form shown in Fig. 4.6. From the analogy with Ohm's Law, the ratio of the body-ambient temperature gradient to the heat loss per unit surface area should give an estimate of thermal insulation per unit area. One point is plotted for each of the 31 pigs exposed to 5°C, taken from the second hour of each run, and the slope of the regression of $A(T_R - T_A)$ on H then gives an index of the mean thermal insulation, since

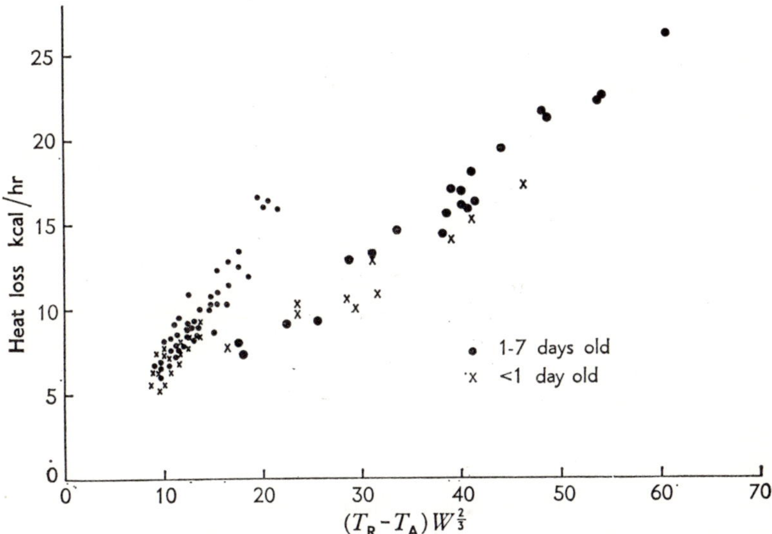

FIG. 4.6. The relation between rate of heat loss from new-born pigs and the product of the rectal-ambient temperature gradient and body weight to the two-thirds power (a measure of the animal's surface area). The larger circles and crosses refer to results obtained at 5°C; the smaller circles and crosses refer to results at 33°C. The regression of $(T_R - T_A)W$ on the heat loss gives an index of the mean thermal insulation at each temperature.

$$\text{thermal insulation} = \frac{T_R - T_A}{H/A} \qquad (1)$$

$$= \frac{A(T_R - T_A)}{H}$$

where $\qquad T_R$ = rectal temperature,

$\qquad\qquad T_A$ = ambient temperature,

$\qquad\qquad H$ = heat loss per unit time, including loss from heat storage,

$\qquad\qquad A$ = animal's surface area.

Corresponding values from experiments carried out at 30°C are also plotted in Fig. 4.6. The effective insulation, from the slope, is much less in the 30°C animals.

This is not the actual thermal insulation since H is the total heat loss and not simply the non-evaporative part. The true insulation value for each animal, at 5 and 30°C and from experiments at other temperatures, has also been calculated from the non-evaporative heat loss per unit area and the temperature gradient. Evaporative heat loss from the new-born pig is so small in quantity that it can all be accounted for, approximately, by respiratory loss (Mount, 1962*a*). For these calculations of insulation all evaporative loss is therefore assumed to take place from the respiratory tract, which acts as a shunt between the body core and the environment. The non-evaporative heat loss is then taken as the appropriate proportion of the total rate of heat production plus rate of heat storage; at 30°C, the proportion is close to 89% (see Chapter 6). An estimate of the heat storage rate is obtained by multiplying the mean rate of change of rectal temperature by body weight and by the assumed specific heat of 0·83 cal/g °C (Burton & Edholm, 1955). The results of the calculation are shown in Fig. 4.7. A significant feature is that the calculated thermal insulation continues to increase as ambient temperature falls at levels below the critical temperature, without an abrupt change in insulation occurring at the critical temperature. The principal change would be expected to occur at the critical temperature if factors in the tissue insulation, particularly peripheral vasoconstriction, were the predominant ones in determining the animal's total insulation. Fig. 4.7 suggests that these factors are not predominant, but before that conclusion can be reached it is necessary to confirm that the critical temperature, which so far has been estimated only from metabolic rate measurements, has been assigned the correct value.

Another way of measuring the critical temperature is to determine at what point on the ambient temperature scale peripheral vasoconstriction, with its accompanying rise in tissue insulation, occurs as conditions become cooler. The critical temperature measured in this way may not always be the same as that estimated as the temperature at which metabolic rate shows a rise above the minimum characteristic of the thermoneutral zone. Thus Cannon & Keatinge (1960) found that although fat men increased their

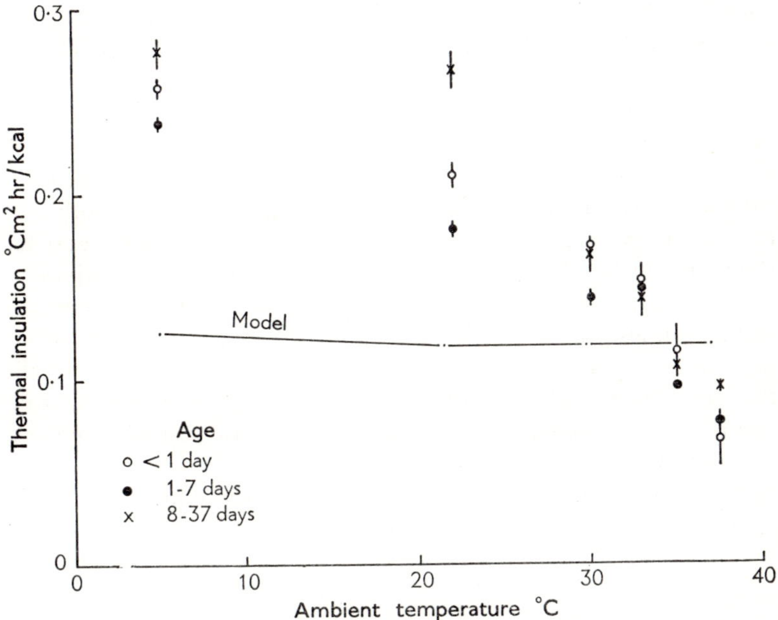

FIG. 4.7. Thermal insulation per unit of surface area (means and standard errors) over a range of environmental temperature for young pigs from birth to 37 days of age (from Mount, 1963*b*, by permission of *Journal of Physiology*).

metabolic rates in water cooled below 33°C, they did not show maximum tissue insulation until the water temperature became much lower, about 12°C. The thinnest men in their experiments, on the other hand, showed a rise in metabolic rate and developed maximum tissue insulation at much the same temperature, close to 33°C. (In air, the critical temperatures would have been lower, about 28°C, on account of the additional insulation provided by the air itself.) Thus in the fat men there were two distinct thresholds, the 'metabolic threshold temperature', and a 'maximal tissue insulation temperature', but in the thin men these two were similar.

The new-born pig would be expected to behave more like a thin man than a fat man, because the animal has no effective subcutaneous fat layer (Widdowson, 1950). In any case, if the two thresholds were to coincide, they would provide independent methods of verifying the critical temperature; if they did not coincide, the maximal tissue insulation temperature would be the value to use in a discussion of thermal insulation, rather than the metabolic threshold.

To estimate the temperature at which peripheral vasoconstriction takes place in the new-born pig (Mount, 1964*a*), the thermal circulation index (Burton & Edholm, 1955) has been used. This is calculated from the simultaneous rectal, skin, and ambient temperatures, and gives the ratio of the surface-air (external) to the rectal-surface (tissue) insulation under steady state conditions. The index rises when the tissue insulation falls relative to external insulation, and when this happens it can be taken as an indication of increased peripheral blood-flow. If, as seems likely, nearly all the new-born pig's evaporative heat loss takes place in the respiratory exchange, then for the purposes of this calculation heat lost from the animal's surface may all be assumed to be in the non-evaporative form. Using the same notation as in eqn. (1):

$$R_E \doteq \frac{T_S - T_A}{H} \tag{2}$$

$$R_I = \frac{T_R - T_S}{H} \tag{3}$$

where R_E is the external insulation, R_I is the tissue insulation, and T_S is the skin temperature. The thermal circulation index is then given by:

$$\frac{R_E}{R_I} = \frac{T_S - T_A}{T_R - T_S} \tag{4}$$

Skin temperatures calculated as the mean of five simultaneous thermocouple values taken from the head, right flank, left flank, middle of the back, and the abdominal surface of the new-born pig are plotted in Fig. 4.8 against ambient temperature. There is no clear-cut change in the relation until the ambient temperature exceeds 35°C, when the slope begins to increase. Quite a different picture (Fig. 4.9) is produced when, instead, the thermal circulation index is plotted against ambient temperature. Above 30°C the index begins to increase then rises steeply above 33–34°C. This method of determining the critical temperature thus gives the same answer as that obtained from measurements of metabolic rate. Skin temperatures measured from the ears and the limbs may give

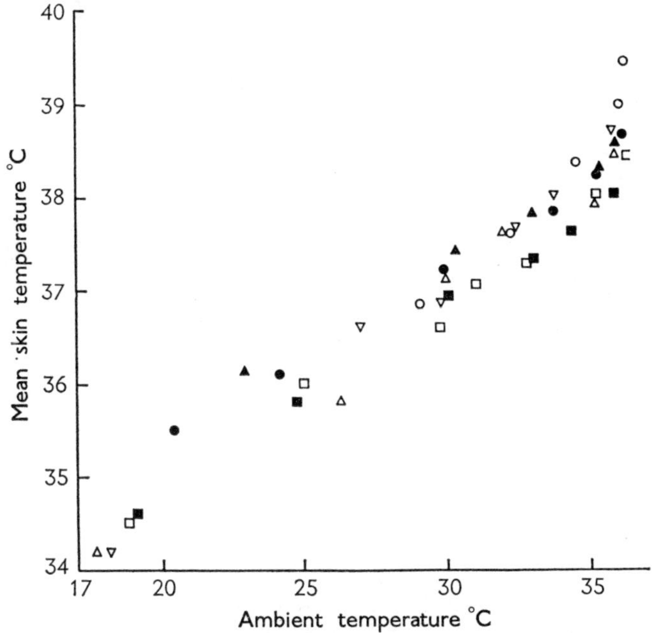

FIG. 4.8. Skin temperature (mean value from head, both flanks, back, and abdomen) of twelve pigs up to 9 days old plotted against environmental temperature.

an even sharper definition of the critical temperature, but mean trunk temperatures are useful in this case because they represent the areas from which most of the animal's heat is lost. Ears and limbs are good indicators of skin temperature change, but if the primary concern is with heat exchange, the rest of the body is quantitatively more important, owing to its larger area, at least in a nearly naked species like the pig.

Now that the new-born pig's critical temperature is established, both for metabolic rate and tissue insulation, it is appropriate to return to the problem posed by Fig. 4.7: why does the thermal insulation rise progressively as the ambient temperature continues to fall below the critical level? The most likely solution to this problem lies in the postural change the animal makes as the temperature falls. In warm conditions, the limbs are stretched out

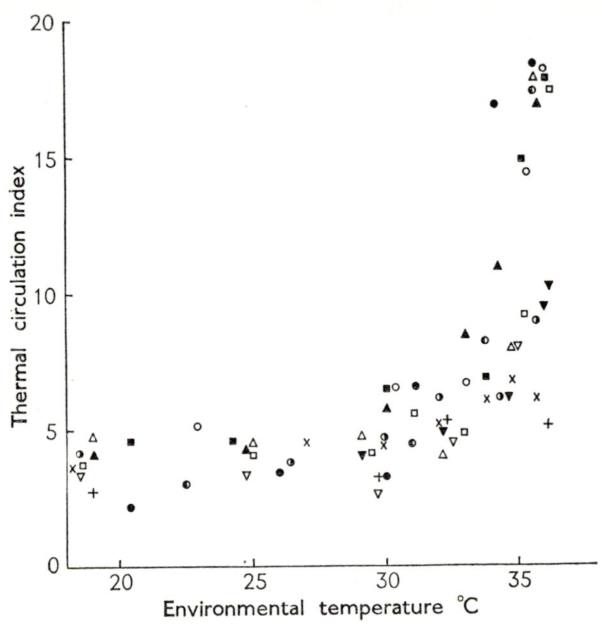

FIG. 4.9. The relation between the thermal circulation index and environmental temperature for the twelve new-born pigs referred to in Fig. 4.8 (from Mount, 1964*a*, by permission of *Journal of Physiology*).

and the pig lies relaxed and extended on its side. In cold conditions, the pig crouches, shivering violently, sometimes with the forelimbs thrust between the hind limbs, and with its sparse coat of hair erected. All grades of postural adjustment between the warm extended and the cold flexed attitudes are seen with intermediate temperatures. What is happening is that in the flexed posture less of the animal's surface is exposed to the environment for heat exchange by radiation and convection than in the extended posture (Mount, 1964*b*, and see Chapter 7). This means that for purposes of heat exchange the *effective* surface area of the animal is reduced in the flexed position, and this leads to an increase in the *effective* overall thermal insulation of the animal. Per unit *effective* surface area, however, the insulation is much the same in the two postures. If the term for area used in arriving at the thermal insulation, °C.m².hr/kcal, is not taken as m² of the *whole* surface area, but only as m² of the *effective* surface area, the result is more nearly that which would be expected if the insulation remained largely unchanged below the critical temperature. Consequently the

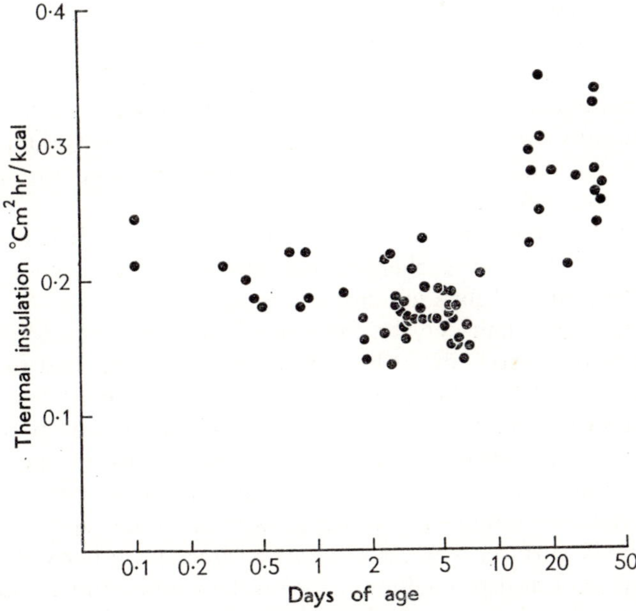

FIG. 4.10. Thermal insulation per unit of surface area for pigs up to 37 days of age at 22°C environmental temperature (from Mount, 1963*b*, by permission of *Journal of Physiology*).

continued rise in thermal insulation is independent of any further vascular change.

The rise in metabolic rate shown by the new-born pig in the post-natal period, and the consequent effects on the relation between metabolic rate, thermal insulation, and the body-ambient temperature gradient, may now be reconsidered. Fig. 4.10 suggests that during the first few days after birth the thermal insulation at

22°C, which is a relatively cool environment for a new-born pig, tends to fall, perhaps in consequence of a relaxation of posture. The post-natal rise in metabolic rate may thus be associated not only with the rise in body temperature discussed earlier, but also with an increased heat loss due to a fall in effective insulation. The increase in thermal insulation after the age of one week, on the other hand, is probably due to the acquisition of a subcutaneous fat layer.

Coat insulation

If the new-born pig's sparse coat of hair does contribute effectively to the animal's thermal insulation, pilo-erection might be expected to increase the surface-air insulation. There is a small but definite insulative effect attributable to the piglet's hair, which at maximum amounts to about 15% of the total insulation. The effect of the coat is not measurable by the time the pig is two weeks old (Mount, 1964a).

Berry and Shanklin (1961) found that a calf with a hair density of 9·8 mg/cm^2 had a thermal insulation of the hair coat of 0·008°C.m^2.hr/kcal for each unit of hair density of 1 mg/cm^2. For the same unit of hair density, the coat of the new-born pig has an insulation of 0·020°C.m^2.hr/kcal. It seems, therefore, that for the same density of hair the pig's hair coat confers a higher degree of thermal insulation than the calf's coat. The comparison is not entirely valid, however, because the actual density of the calf's coat is greater than that of the pig's coat, and because the values for the calf were obtained by direct measurement of heat flow made at the body surface, whereas the piglet values were derived from estimations of heat production made on the whole animal. Barnett (1959) measured the thermal insulation of isolated mouse skin, shaved and unshaved. Calculation from his results shows the insulation to be 0·080°C.m^2.hr/kcal for a unit of hair density of 1 mg/cm^2, an insulation four times that of pig hair at the same density. The actual hair densities of pig and mouse were 1·5 and 0·6 mg/cm^2 respectively, and consequently were more comparable with each other than with the calf. The greater insulation provided by the mouse's coat may be due to finer hairs producing, weight for weight, a more effective insulating layer of trapped air. The measurement of density as number of hairs per unit area may provide a more useful basis for comparisons than the measurement of density as weight of hair per unit area.

METABOLIC CAPABILITY

One of the factors which would be important in determining whether an individual pig in the cold manages to maintain its body temperature or not is the maximum rate of heat production which can be sustained by the animal. Such a rate is a measure of the level of 'metabolic capability' (Mount, 1961). This is distinct from the highest metabolic rate measurable, which can be only briefly sustained and so bears little thermoregulatory significance.

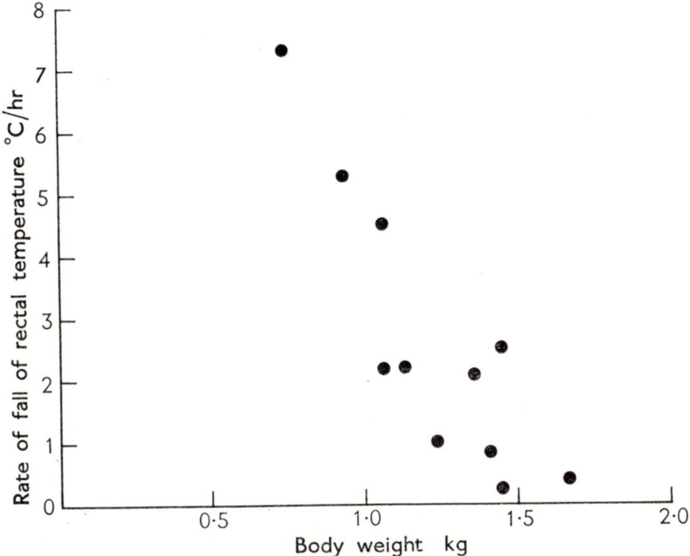

FIG. 4.11. The mean rates of fall of rectal temperature in eleven pigs under one day old, exposed to 5°C environmental temperature, in relation to body weight (from Mount, 1967*b*, by permission of Lea & Febiger).

What is important for thermoregulation is measurement of the response of the organism to a steady long-lasting stimulus, such as was provided by the experiments on pigs at 5°C mentioned earlier. The effects on body temperature in these experiments are shown in Fig. 4.5, and suggest that comparisons between animals should be made at a given body temperature, if they are to be realistic. When, in spite of a vigorous metabolic response, the rectal temperature is falling in a cold environment, it is likely that the maximum

metabolic rate is reached. However, the maximum rate, or metabolic capability, is limited by the body temperature of the organism in which it is occurring: maximal metabolic capacity of tissue
necessarily falls as tissue temperature falls. Fig. 4.5 suggests that
5°C represents an approximate cold limit for the single new-born
pig, and that it provides suitable conditions for discriminating
between pigs in respect of thermoregulatory ability, which may be
measured either in terms of the rate of fall of rectal temperature,
or in terms of the reciprocal expression, cold-resistance, which can

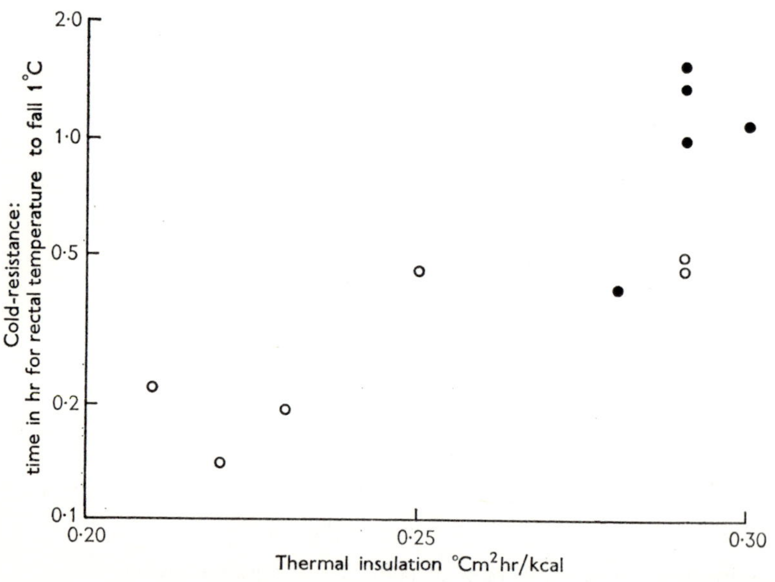

FIG. 4.12. Cold resistance, estimated as time for rectal temperature
to fall 1°C, plotted against thermal insulation, for the eleven pigs at
5°C referred to in Fig. 4.11. ○, body weight less than 1.2 kg; ●,
body weight above 1.2 kg.

be arbitrarily taken as time in hours for rectal temperature to fall
1°C. There were eleven pigs under one day of age in the group of
31 pigs at 5°C, and the mean rates of fall in rectal temperature for
the under-one-day pigs are plotted against body weight in Fig.
4.11. The heavier pigs show lower rates of fall, and the intercept
on the body weight axis is approximately 1·5 kg. This suggests that
a pig weighing 1·5 kg and more at birth and exposed to 5°C would

show only a small fall, if any, in rectal temperature, a finding which is in accordance with the general husbandry experience that new-born pigs of this weight have a higher degree of cold-resistance than lighter pigs. The fundamental dependence of cold-resistance on overall thermal insulation is shown by Fig. 4.12, which also demonstrates the partial dependence of thermal insulation itself on body weight. Even with these correlations, however, a statistical analysis of the rates of fall of body temperature in new-born pigs showed that body weight and age together

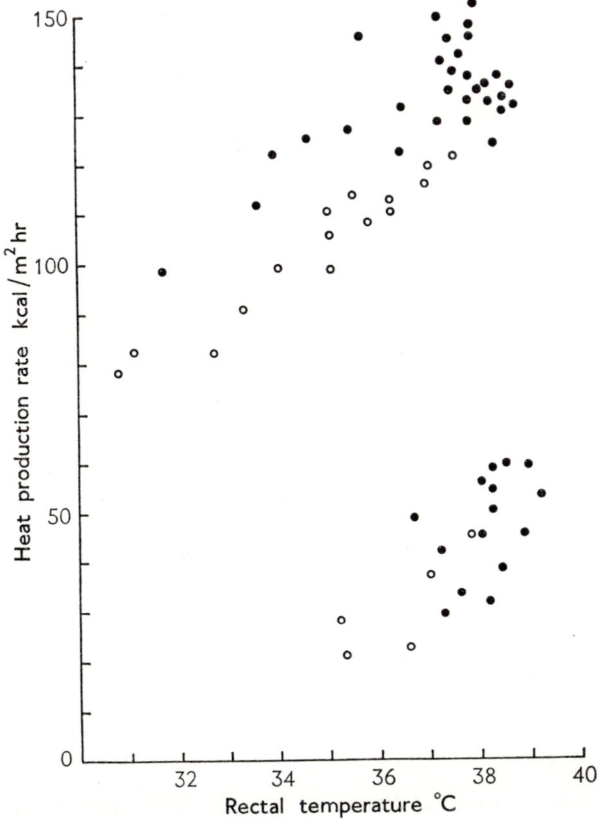

FIG. 4.13. Heat production rates for new-born pigs, at values approaching the maximal at 5 °C environmental temperature (upper points), and at 33 °C, near thermal neutrality (lower points), plotted against rectal temperature. Open circles: under one day old; solid circles: 1–7 days old (from Mount, 1966*b*, by permission of Frayn Printing Company, Seattle).

accounted for no more than about two-thirds of the variance associated with rectal temperature change, indicating that additional factors are important in determining the response to cold.

The metabolic rates of the 5°C pigs are plotted against rectal temperature in Fig. 4.13. The values given are those for the second, and, where the experiment was not terminated earlier by marked hypothermia, the fifth hour of exposure. Two things are apparent from Fig. 4.13: the first is the relation between metabolic rate and

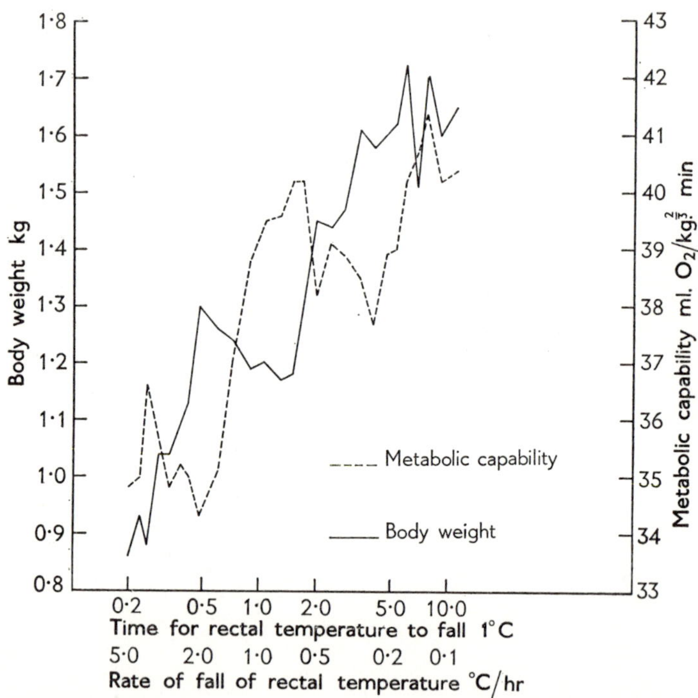

FIG. 4.14. Moving average curves relating the body weight and metabolic capability of 31 new-born pigs, exposed to 5 °C, to the time for rectal temperature to fall 1 °C (cold resistance).

rectal temperature, and the second is the lower metabolic rate of pigs less than one day old when compared with older pigs which had the same rectal temperature. Not only, therefore, do pigs in the immediately post-natal period have lower metabolic rates than slightly older pigs at the same ambient temperature; the metabolic rates are also lower at the same rectal temperature, indicating a rise in the pig's metabolic capability during a period of post-natal change.

Metabolic capabilities were therefore estimated for a given level of rectal temperature. For this purpose 36°C was chosen, since measurements in the region of this rectal temperature were obtained from a majority of the animals used in the experiments. Even so, in some cases extrapolation was required; this was carried out by assuming a Q_{10} of 2, a value which appears from the 5°C results in Fig. 4.13 to be appropriate.

Fig. 4.14 shows moving average curves relating body weight and metabolic capability to cold-resistance. The striking feature is the apparent alternation between body weight and metabolic capability so that at a given level of cold-resistance a low body weight is compensated by a higher metabolic capability. The product of body weight and metabolic capability is plotted against cold resistance in Fig. 4.15. The clear correlation which results, in

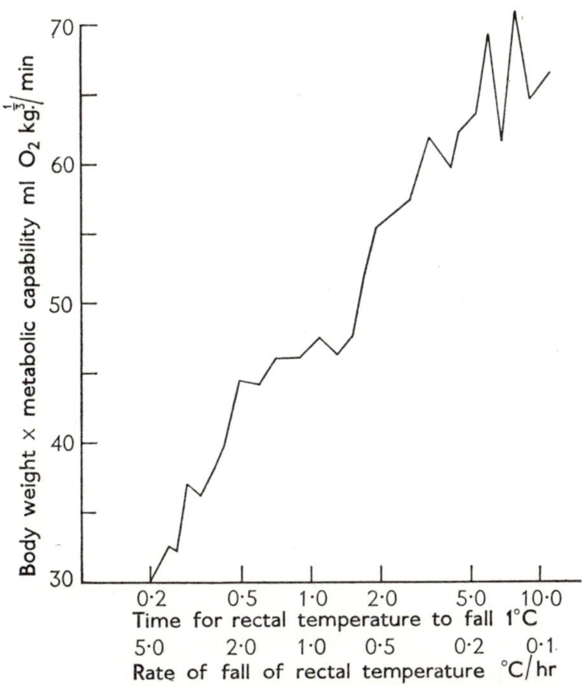

FIG. 4.15. Moving average curves of the product of body weight and metabolic capability plotted against time for rectal temperature to fall 1°C (cold resistance), from the results of Fig. 4.14.

contrast with that obtained between body weight alone and cold resistance, indicates that, in addition to body weight and age, metabolic capability must be considered as an important factor in determining cold-resistance.

Measurements of oxygen consumption rate were also made at an environmental temperature of 33°C, close to the new-born pig's critical temperature. The results are included in Fig. 4.13, and represent minimum metabolic rates for corresponding rectal temperatures. The ratio between maximum and minimum rates, which Giaja (1925) has termed the 'metabolic quotient', lies between 3 and 4, in common with a number of other species (see table on p. 284 of Brody, 1945).

THE RELATION BETWEEN METABOLIC RATE AND RECTAL TEMPERATURE

Fig. 4.13 suggests that under some conditions there is a relation between an animal's metabolic rate and its rectal temperature. The Q_{10} is the factor by which the velocity of a reaction is increased for a rise in temperature of 10°C, and is given by the equation:

$$Q_{10} = \left(\frac{K_2}{K_1}\right)^{10/(T_2 - T_1)}$$

where K_1 and K_2 are velocity constants corresponding to temperatures T_1 and T_2.

As an example of the variation of reaction velocity with absolute temperature in animals, Prosser (1950) reasons that the frog swimming in ice water cannot possibly go faster than energy becomes available in its muscles. The same argument applies in the homeotherm, and is the basis for considering that metabolic rate is limited by body temperature in the case of the pigs at 5°C environmental temperature. A rate–temperature relation exists under conditions where metabolic rate is either at an uncontrolled maximum, or at an uncontrolled minimum. For all values between the maximum and minimum, thermoregulatory mechanisms operate, and then there is no reason to expect any relation between metabolic rate and body temperature. In the pigs in the cold at 5°C, some at least of the animals were producing heat at the maximum rate; no thermoregulatory control mechanisms could operate, and it might be expected that the heat production rate would be a function of the body temperature. Fig. 4.13 does suggest a Q_{10} of about 2, a common value for chemical reactions. In the warm, a similar value has been found for the sheep at ambient temperatures above the critical level (Graham, Wainman, Blaxter & Armstrong, 1959); under such warm conditions, where metabolic rate is either minimal or rising passively with body temperature, there are again no control mechanisms operating.

Another situation in which thermoregulatory control is absent or deficient occurs in the anencephalic human infant. Cross, Gustavson, Hill & Robinson (1966) measured the rate of oxygen consumption in such an infant, and found a characteristic rate-temperature

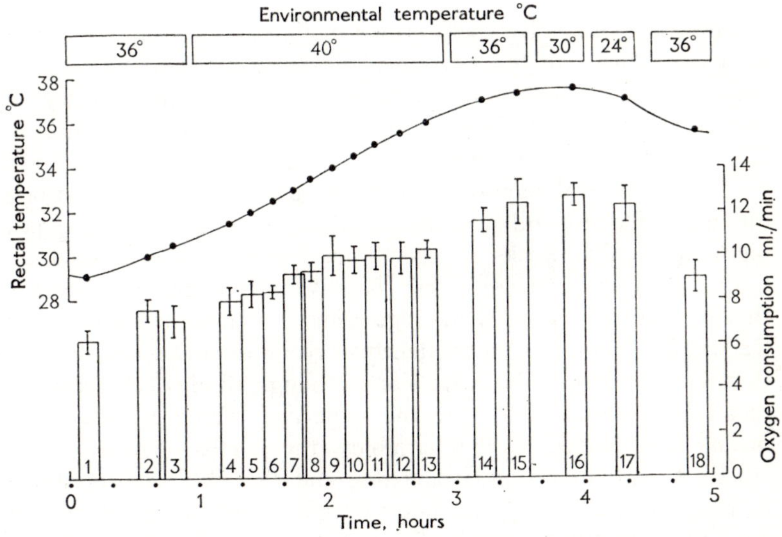

FIG. 4.16. Rectal temperature (continuous line) and oxygen consumption (blocks) of an anencephalic infant exposed to the environmental temperatures given at the top of the diagram. Each oxygen consumption block is shown with ±90% confidence limits (1·8 × S.E. mean); each block is numbered in the sequence in which the measurement was made, and a solid circle on the rectal temperature plot corresponds to each block (from Cross, Gustavson, Hill & Robinson, 1966, by permission of *Clinical Science*).

relation over the whole temperature range tested (Fig. 4.16), just as in poikilotherms (Valen, 1958; Fuhrman & Fuhrman, 1959). Normal babies, on the other hand, show no apparent relation between metabolic rate and rectal temperature (Adamsons, Gandy & James, 1965), and would not be expected to, since any changes in metabolic rate in the intermediate zone are not a function of the level of body temperature, but are a function of the control mechanism. In the normal new-born infant, which responds to cooling by increasing its metabolism, it is only at the hot and cold extremes, where no control can operate, that body temperature determines metabolic rate; the rate-temperature relation then becomes apparent, as in the case of new-born pigs illustrated in Fig. 4.13.

AGE AND BODY WEIGHT

The effect of age by itself on metabolic rate does not appear to be a large factor in the new-born pig beyond the first few days post-partum. To determine whether this is so, and to what degree the animal's metabolic rate depends on its body weight, Mount & Rowell (1960*a*, *b*) investigated the relation between age, body weight, and rectal temperature on the one hand, and metabolic rate measured at 4 and 30°C on the other, in a total of 98 pigs from birth up to five weeks of age. The relation was examined by multiple regression techniques; the advantage of this approach is that it is possible to study the regression of metabolic rate on body weight, for example, independently of rectal temperature and age; that is, to determine the regression as though all pigs had the same rectal temperature and were the same age. Table 4.1 gives the means and ranges of the variables, the partial regression co-efficients and their standard errors, and the resulting equations relating metabolic rate and body weight, rectal temperature and age. The equations take the form shown since log (oxygen con-sumption rate), log (body weight), rectal temperature, and age were functions used in the analysis. Log (oxygen consumption rate) and log (body weight) were used because the relation between these two quantities tends to follow a power law, of which pro-portionality of metabolic rate to surface area (or the two-thirds power of body weight) is a special case. Rectal temperature itself was used rather than the logarithm since an exponential relation to metabolic rate might be expected. Age was chosen instead of the logarithm simply because age gave a rather smaller residual variation than log (age), and there were no other grounds on which to base a choice between the two.

TABLE 4.1

(a) Means and ranges of oxygen consumption rate, body weight, rectal temperature, and age for pigs in the first 5 weeks after birth

Ambient temper- ature and age	No. of pigs	Oxygen consumption rate (ml./min)	Body weight (kg)	Rectal temperature (°C)	Age
30°C first week	61	30·2 (15·7–55·2)	1·69 (0·95–3·37)	39·1 (37·8–39·8)	2·79 days (5 hr–6·25 days)
4°C first week	18	58·3 (32·5–93·8)	1·58 (1·07–2·72)	38·4 (36·5–39·4)	3·32 days (1 hr–6·83 days)
30°C 1–5 weeks	19	54·4 (29·9–70·5)	4·79 (2·88–7·60)	39·5 (38·6–40·2)	21·32 days (10–37 days)

TABLE 4.1

(b) Partial regression coefficients, and their standard errors, for log (oxygen consumption rate, ml./min) on log (body weight, kg), rectal temperature (°C) and age (hr)

Ambient temperature and age	No. of pigs	Log (body weight, kg)	Rectal temperature (°C)	Age (hr)
30°C first week	61	0·610 ± 0·064	0·157 ± 0·035	0·000748 ± 0·000403
4°C first week	18	0·562 ± 0·136	0·105 ± 0·029	0·001106 ± 0·000763
30°C 1–5 weeks	19	0·684 ± 0·083	0·150 ± 0·038	−0·000372 ± 0·000126

TABLE 4.1

(c) Equations relating oxygen consumption (y, ml./min) to body weight (x_1, kg) rectal temperature (x_2, °C) and age (x_3, hr)

30°C first week: $y = 0\cdot0442 x_1{}^{0\cdot610}\, e^{0\cdot157 x_2}\, e^{0\cdot000748 x_3}$

4°C first week: $y = 0\cdot7134 x_1{}^{0\cdot562}\, e^{0\cdot105 x_2}\, e^{0\cdot001106 x_3}$

30°C 1–5 weeks: $y = 0\cdot0598 x_1{}^{0\cdot684}\, e^{0\cdot150 x_2}\, e^{-0\cdot000372 x_3}$

From Mount & Rowell, 1960b, by permission of *Journal of Physiology*.

The results of the analysis show that in the first five weeks after birth, body weight and rectal temperature both had a large effect on consumption. The effect of age, however, was less striking, and, taking the first week as a whole, it was not certain that it had any

D

effect at all. If the effect of age was in fact confined to the first day or two after birth, this is what would be expected, particularly since metabolic rate and age were negatively related from one to five weeks of age. The negative coefficient for the older pigs does not mean that metabolic rate decreases as the pig grows older; indeed it increases because the pig gets bigger. It does indicate, however, that amongst a number of pigs with the same body weight and rectal temperature, the older pigs will have lower metabolic rates, perhaps because their thermal insulation is higher as the result of more subcutaneous fat.

TABLE 4.2

(*a*) Effects of differences in body-weight on oxygen consumption of pigs at a given age

	Age (*constant*)	Body-weight (*kg*)	Oxygen consumption (*ml./min*)	Difference
30°C first week	3 days	1·50	27·6	3·9
		1·86	31·5	
4°C first week	3 days	1·50	58·5	7·4
		1·86	65·9	
30°C 1–5 weeks	3 weeks	3·83	43·6	13·4
		5·68	57·0	

TABLE 4.2

(*b*) Effects of differences in age on oxygen consumption of pigs of a given body-weight; the age differences used correspond to the weight differences in Table (*a*)

	Body-weight (*constant*)	Age	Oxygen consumption (*ml./min*)	Differences
30°C first week	1·68 kg	2 days	29·0	1·1
		4 days	30·1	
4°C first week	1·68 kg	2 days	60·7	3·3
		4 days	64·0	
30°C 1–5 weeks	4·75 kg	2 weeks	53·8	−6·4
		4 weeks	47·4	

From Mount & Rowell, 1960*a*, by permission of *Nature*.

The relatively small effect of age by itself on metabolic rate as compared with the effect of body weight is demonstrated in Table 4.2. Age is held constant (statistically speaking) at 3 days or 3 weeks in Table 4.2*a*, and body weight varied; while in Table 4.2*b* body weight is held constant and age varied. The relative variations are realistic since they are taken from the mean curve of body weight on age. The differences in metabolic rate produced by varying body weights are decidedly higher than those produced by varying age; in the three-week pig the difference is positive for a weight increase and negative for the corresponding age increase.

With this clear evidence of the effect of age on metabolic rate being confined to the first day or two following birth, and the negligible mean effect of age during the first week, it was of interest to find Holub, Forman & Ježková (1957) concluding that chemical thermoregulation in pigs continued to develop up to 20 days following birth. Their results are given in Table 4.3. They found a fall in oxygen consumption at 3°C ambient temperature (30 min exposure) as compared with 23°C up to six days of age; after this age, metabolic rate increased at 3°C. This is quite contrary to the findings discussed earlier in this chapter, and for this reason the experiments of Holub *et al.* were repeated as closely as possible (Mount, 1958, 1959); the results are given in Table 4.4. There is no doubt that under the standard conditions used, in a calorimeter chamber with low air speed, the oxygen consumption rate of pigs under six days of age was definitely increased at 3°C as compared with 23°C ambient temperature. The difference between the two sets of results may be partly accounted for by the fact that Holub *et al.* used a mask technique for oxygen consumption, with the piglet exposed in a cold room (A. Holub, personal communication). Under these conditions, the animal's micro-environment would probably exert a higher thermal demand, due to a higher rate of air movement at the pig's surface, even though the air temperature used in the two sets of experiments was comparable. Another possibility is that the breeds of pig used had different susceptibilities to cold. These factors could account for the two sets of results, particularly since 3°C would be in the region of the new-born pig's cold limit, as discussed earlier. Amongst the 31 pigs at 5°C, however, even the most rapidly failing animal showed a marked elevation in metabolic rate during the first hour of exposure, which would cover the period of Holub's observations. In respect of the statement by Holub *et al.* that chemical thermo-regulation is age-dependent up to 20 days, and that age is a more important factor than weight, their analysis does not allow a distinction to be made between the effects of age and those of body

TABLE 4.3

The results of Holub *et al.* (1957) relating to the
metabolic rates and body temperatures of new-born
pigs exposed to cold

Age (days)	Environmental temperature, °C	Oxygen consumption rate, ml./kg. min
3	23 ± 1	20·2 ± 3·1
	3 ± 1	15·9 ± 5·1
4–5	23 ± 1	19·2 ± 8·3
	3 ± 1	17·4 ± 7·9
6	23 ± 1	16·4 ± 3·6
	3 ± 1	16·4 ± 2·9
9–10	23 ± 1	14·9 ± 2·7
	3 ± 1	23·3 ± 4·6

	Decrease in body temperature in °C in 30 min at 3°C; in further 30 min at 23°C	
3	1·8 ± 0·14	0·8 ± 0·17
6	1·8 ± 0·25	0·8 ± 0·18
9–10	1·2 ± 0·19	0·6 ± 0·11

weight, so that the effects ascribed to age could have in fact been
due to body weight increase. It is relevant that A. B. Cairnie
(1958, personal communication), using a gradient layer calorimeter,
confirmed Mount's findings of a rise in metabolic rate in the cold
in new-born piglets.

COMPARISON WITH OTHER SPECIES

Human infant

The new-born pig, weighing rather more than 1 kg, is close
enough in size to the 3 kg human infant for useful comparisons to
be made. In both species, the thermal insulation of an effective
coat is lacking, a weakness in the face of cold which is offset by the
pig huddling with its litter-mates in a nest, while the baby is
clothed, and shielded from the environment in other ways. The
levels of maturity can be measured in terms of the eyes being open,
and shivering taking place in the cold. The new-born pig, with its
capacity for movement over the ground from birth, is capable of
greater variation of response to environment than is possible for the
immobile baby. The rates of growth are quite different: whereas

TABLE 4.4

Oxygen consumption and body temperature in new-born pigs exposed to the cold (from Mount, 1958, 1959)

(*a*) Oxygen consumptions of new-born Large White pigs at two temperatures near 23 and 4°C; in each case the temperature sequence was 23–4–23. Both temperature and consumption are given as means with standard errors (6 animals in each age group).

Age (days)	Environmental temperature (°C)	Oxygen consumption (ml. dry gas s.t.p./kg.min)
Less than 1	22·8 ±0·4	21·6 ±1·1
	3·1 ±0·1	34·5 ±1·1
	22·2 ±0·1	24·9 ±1·0
2–4	23·3 ±0·9	25·7 ±0·9
	3·8 ±0·9	38·8 ±0·9
	23·0 ±0·2	26·0 ±0·6
5–6	24·1 ±0·6	22·7 ±0·6
	5·0 ±0·5	35·9 ±1·2
	23·6 ±0·5	22·9 ±1·0

(*b*) Initial rectal temperatures of new-born pigs and changes occurring during each period of measurement. Six pigs in each age group.

Age	Initial rectal temperature (°C) Mean ± S.E.	Rectal temperature changes (°C) during 1st period (23°C) Mean ± S.E.	2nd period (4°C) Mean ± S.E.	3rd period (23°C) Mean ± S.E.
Under 1 day	38·6 ±0·1	−0·4 ±0·1	−1·7 ±0·3	0·9 ±0·2
2–4 days	39·1 ±0·2	−0·3 ±0·02	−0·6 ±0·2	0·2 ±0·2
5–6 days	39·5 ±0·3	−0·4 ±0·1	[ ]0 ±0·1	−0·2 ±0·1

the piglet doubles its birth weight in one week, the baby takes six months to achieve the same relative increase. To reach mature size, however, the pig has farther to go, since at birth it is only about 1% of the mature adult size, whereas the human infant is about 5% of adult size at birth.

At birth both the pig and the baby show a fall in body temperature from which recovery takes place more rapidly in the pig. The level of rectal temperature in the mature pig is in the vicinity of 39°C, whereas in man it is 37°C. This complicates direct comparisons of metabolic rate in the two species, since the same environmental temperature imposes different temperature gradients, and therefore different heat flows to the environment, in the two cases. Another difficulty in the comparison arises from the different body weights: these produce different metabolic body sizes (see Chapter 11).

Initially, a comparison can be made as in Fig. 4.17. This shows some results from human infants superimposed on the curves for pigs from Fig. 4.2c relating oxygen consumption rates to body

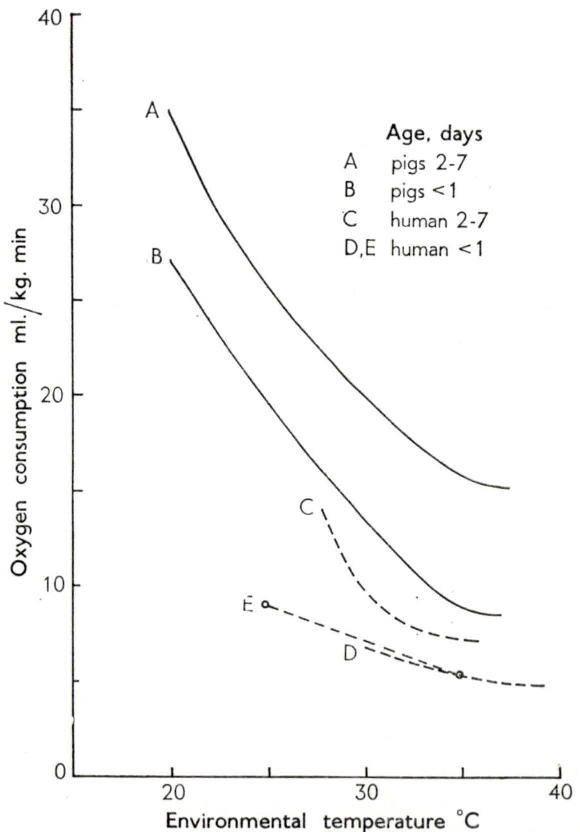

FIG. 4.17. Rates of oxygen consumption per kg for new-born pigs and human infants over a range of environmental temperature. A and B from Mount (1959); C and D from Hill & Rahimtulla (1965); E from Adamsons, Gandy & James (1965) (from Mount, 1966b, by permission of Frayn Printing Company, Seattle).

weight and environmental temperature. The levels in the two species are different although the pattern with age appears to be similar, and critical temperature is high at birth in both. Hill & Rahimtulla (1965) produced clear evidence of a rising basal metabolism in the human infant in the days following birth. In Fig. 4.18 the same measurements are presented as ml.O_2/kg 0·75 min, which largely removes the effects of differing body size, against an environmental temperature scale which is made relative to rectal temperature. This enhances the similarity between the two species.

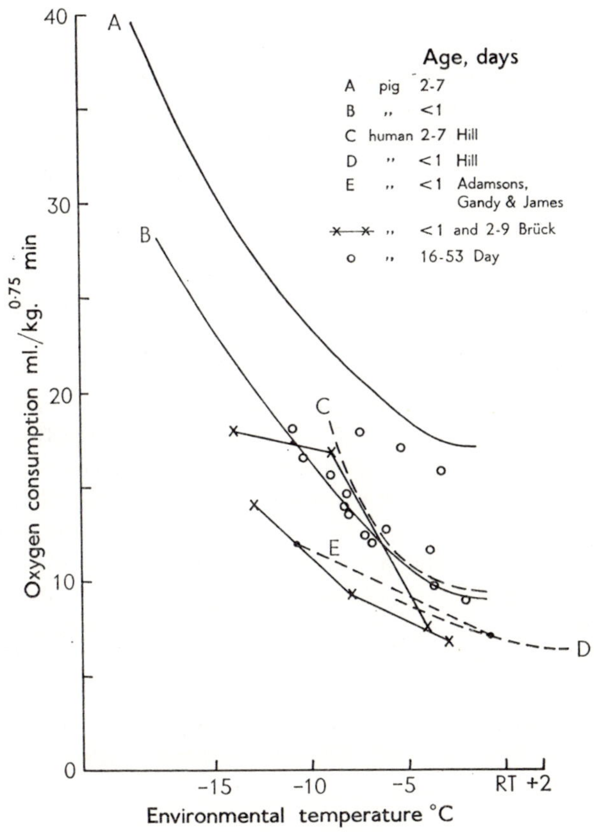

FIG. 4.18. The results of Fig. 4.17 recalculated per kg$^{0.75}$ and plotted on an environmental temperature scale made relative to rectal temperature.

Lamb

The new-born lamb, at 4 kg body weight, is larger than the human infant and much larger than the new-born pig. Metabolic rate rises soon after birth, but this is associated with the intake of milk; age has little effect on metabolic rate. The type of coat influences the lamb's thermal insulation, which is in any case considerably higher than that of the pig (Alexander, 1961*b*). The overall insulation at − 10°C ambient temperature is 0·40°C.m^2.hr/kcal for lambs with fine coats, and 0·54 for lambs with hairy coats; these values may be compared with 0·27 for the new-born pig at 5°C ambient temperature.

Another way in which the lamb differs from the pig is in the lamb's higher metabolic capability, or 'summit metabolism' (Alexander, 1962). This is about 250 kcal/m^2.hr, compared with the highest single value found in the new-born pig of 162 kcal/m^2.hr, and a mean value in the pig of 133. The combination of a higher metabolic capability and a higher thermal insulation confer a greater degree of cold-resistance on the lamb.

Other species

The rise in metabolic rate from the level found in the period following birth appears to be a general phenomenon (Taylor, 1960). It occurs not only in pig, sheep, and man, but also in the monkey (Dawes, Jacobson, Mott & Shelley, 1960), and in the puppy (Gelineo, 1954; McIntyre & Ederstrom, 1958), although the puppy shows a rising metabolic rate over several days. The development of heat conservation mechanisms lags behind the development of metabolic rate in the puppy, and this is true of the pig, since although the pig shows peripheral vasoconstriction in the cold from birth, it is only after some days that its subcutaneous fat has reached a level which allows the insulative value of vasoconstriction to become effective. The calf (Roy, Huffman & Reineke, 1957), however, and the lamb, have by virtue of their coats an insulative advantage in the development of homeothermy over the less well insulated new-born of other species.

CATECHOLAMINES AND METABOLIC RATE

Moore & Underwood (1960) showed that the subcutaneous injection of noradrenaline in the new-born kitten produced a marked increase in metabolic rate, whereas adrenaline at the same dose level was relatively inactive. A similar stimulation of metabolism in the new-born animal following the injection of noradrenaline was observed in the rabbit and rat (Moore & Underwood, 1963), and in the human infant (Karlberg, Moore & Oliver, 1962). Subsequent work on the new-born rabbit showed the probable relation between the effect of noradrenaline and the activity of brown fat. Local heat production in brown fat is increased by the infusion of noradrenaline (Dawkins & Hull, 1964); when the brown fat is excised, the rise in oxygen consumption during infusion is greatly reduced (Hull & Segall, 1964).

The new-born pig appears to have no brown fat; as pointed out earlier, its total fat content is only about 1% of the body at birth.

The subcutaneous injection of noradrenaline and adrenaline in a dose of 600μg/kg gives the results on rate of oxygen consumption shown in Fig. 4.19. Noradrenaline had no effect during the first

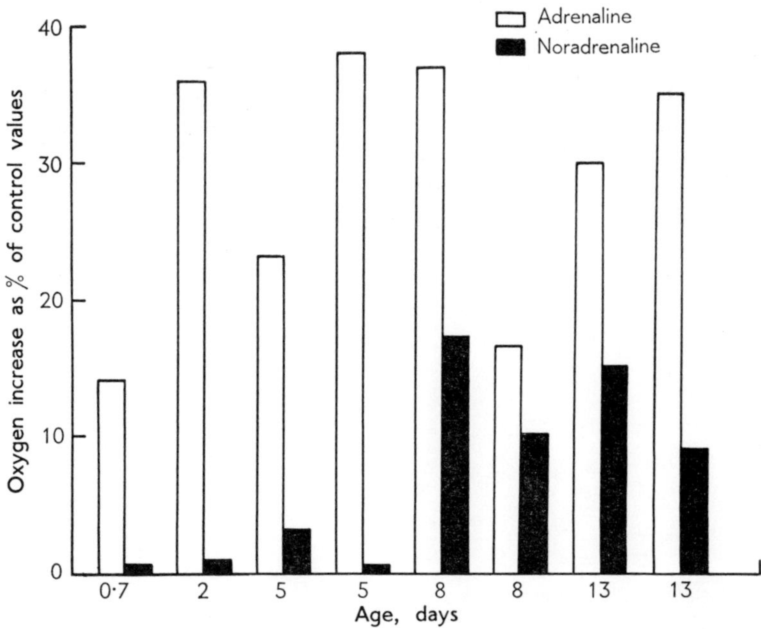

FIG. 4.19. The effects of adrenaline and noradrenaline, in doses of 600μg per kg subcutaneously, on the rate of oxygen consumption in 19 pigs up to 13 days of age (from LeBlanc & Mount, 1968, by permission of *Nature*).

week after birth, and a relatively small effect during the second week, whereas adrenaline had marked effects in both weeks. That this was not due to inactivation of noradrenaline is shown by the marked pressor effect of noradrenaline in the new-born pig (LeBlanc & Mount, 1968).

The pig is thus different from the new-born of some other species in this respect, since in the kitten, rabbit, and rat noradrenaline acts as a metabolic stimulant. The newly hatched chick, however, shows only a slight increase in oxygen consumption rate with noradrenaline, while the effect of adrenaline is more marked (Freeman, 1966); the chick and the pig, therefore, have this much in common.

THYROID

Slebodzinski (1965*a*, *b*) found the level of free thyroxine to be highest in the first post-natal day in the new-born pig, and then to reach a relatively constant level between three and six days later. Since only unbound thyroxine can be considered as the active form of the hormone, Slebodzinski considers the high concentration following birth to be important in post-natal metabolic adaptation. Following birth, there is a considerable increase in the binding capacity in the plasma proteins which determines the level of free, unbound active hormone. Equilibrium between the plasma proteins and thyroid hormone is reached by about eight days of age.

5

ADAPTATION TO THE THERMAL ENVIRONMENT IN THE GROWING PIG

As the pig increases in size, the critical temperature falls progressively from the value of 34°C, which is characteristic of the new-born pig, until at three months of age it is about 25°C; the animal becomes less sensitive to cold, and, later, more susceptible

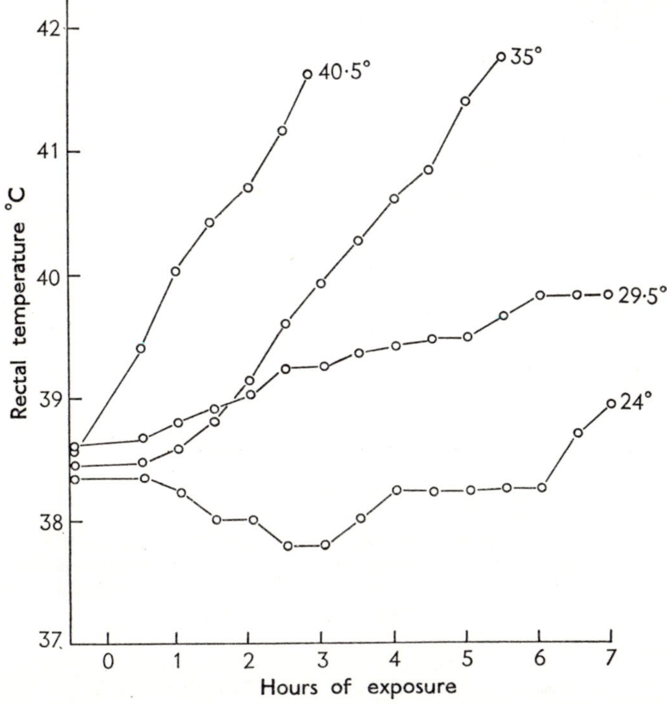

FIG. 5.1. Mean reactions of the rectal temperatures of three male Berkshire pigs, each of about 60 kg body weight, to environmental temperatures of 24, 29·5, 35, and 40·5°C, at a constant relative humidity of 65% (from Robinson & Lee, 1941, by permission of the Royal Society of Queensland).

to heat (see Figs. 5.1 and 5.2). Although the effects of environmental temperature on metabolic rate are then less marked than in the new-born pig, the results are in the same sense.

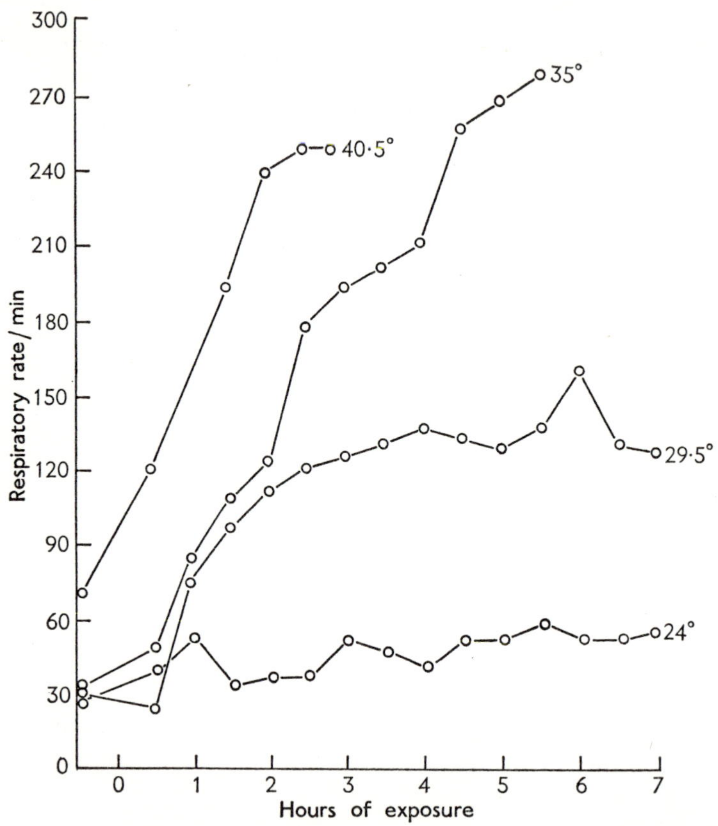

FIG. 5.2. Mean reactions of the respiratory rates of the pigs referred to in Fig. 5.1. during their exposure to different temperatures (from Robinson & Lee, 1941, by permission of the Royal Society of Queensland).

EARLY MEASUREMENTS OF METABOLIC RATE

Some of the earlier metabolic measurements on pigs were concerned with the determination of fasting metabolism. Meissl (1886) found values of about 2300 kcal per 24 hr for the 118 kg pig fasted for 72 hours; Tangl (1912) made measurements on four pigs, two

Yorkshires weighing 40 and 50 kg at 7 months old and two Hungarian pigs weighing 110 to 120 kg at 18 months of age (the younger pigs were much lighter in weight for their age than moddern pigs). Tangl's experiments began after 3 to 4 days' starvation; he measured respiratory exchanges at various temperatures. His results are given in Table 5.1, and suggest critical temperatures of

TABLE 5.1

Mean metabolic rates of pigs of 50 and 100 kg taken from Tangl (1912)

Body weight, kg	Environmental temperature, °C	Metabolic rate kcal/kg.24 hr	Derived critical temperature, °C
50	13–14	33·4	
	17–18	34·1	
	20	28·3	20–23
	23	26·2	
	26	26·7	
100	16–17	19·9	17
	22	19·3	
	26	22·3	

20–23°C for the 50 kg pig, and 17–22°C for the 100 kg pig. Capstick & Wood (1922) used a fasted 10 month pig, and found a critical temperature in the region of 21°C; Deighton (1924, 1929, 1932, 1935) also made observations on metabolic rates in pigs. Ritzman & Colovos (1941) found that when pigs were fasted the metabolic rate fell to about 2600 kcal per 24 hr, and the respiratory quotient to 0·72, after a period of 48 hr.

BODY SIZE, SINGLE PIGS, AND GROUPS

The comparison of the effects of environment on metabolism in animals differing in size and in other respects is complicated by a number of factors, including level of nutrition, association with other animals, and adaptation to particular environments. As the animal becomes larger, so its heat production expressed per unit of body weight falls. This is true for all mammals once they have reached the mature level of heat production, a level which is usually reached by the second post-natal day in the pig. The environment in which the measurement is made also influences the result; in a cold environment a larger animal may well produce more heat per kg.hr than a smaller animal in warm surroundings. The statement about body size and heat production must therefore have regard for the environment, the commonest way of controlling this being to compare metabolic rates at thermal neutrality.

A metabolic rate at thermal neutrality, and under resting conditions, is not the 'basal' rate unless the animal is in the post-absorptive state. If it has eaten recently its metabolic rate will be

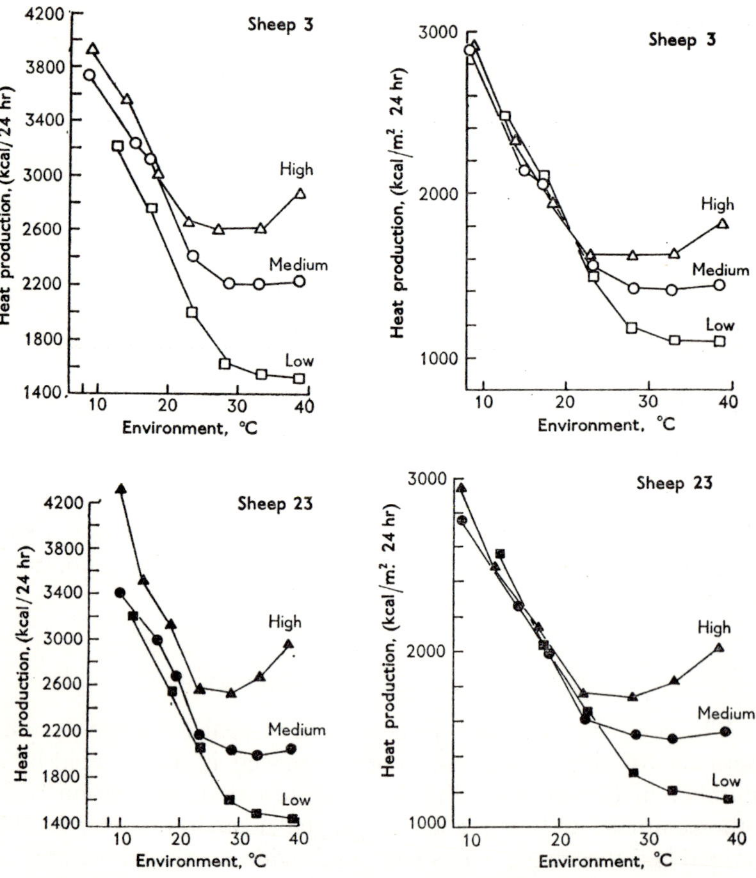

FIG. 5.3. The rate of heat production of two sheep at three feeding levels in relation to environmental temperature. In the diagrams on the left, heat production is given per sheep; on the right, heat production is expressed per square metre of body surface (from Graham, Wainman, Blaxter & Armstrong, 1959, by permission of *Journal of Agricultural Science*).

higher; if it is on a high plane of nutrition its rate will be higher than that on a low plane of nutrition, and the critical temperature will be lower. Blaxter and his colleagues have made many determinations of the relation between metabolic rate, environmental temperature, and level of feeding in the sheep; some of the results are shown in Fig. 5.3. At the low feeding level, metabolism under warm conditions is least, and the critical temperature is highest. As the plane of nutrition is raised, so the thermoneutral level of oxygen consumption increases, and associated with this the critical temperature falls. The relation between thermoneutral rate of heat

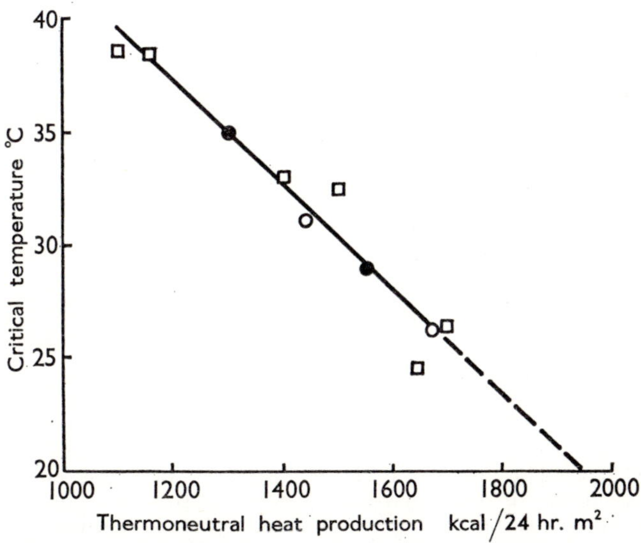

FIG. 5.4. Critical temperature of shorn sheep and thermoneutral heat production. The points refer to wethers (□), non-pregnant ewes (●), and pregnant ewes (○) (from Graham, 1964, by permission of *Australian Journal of Agricultural Research*).

production and critical temperature in shorn sheep is shown in Fig. 5.4. At the lower environmental temperatures, however, heat production becomes independent of feeding level, and is determined instead by the rate of heat loss to the surroundings.

Fig. 5.3 also demonstrates the uniformity which is achieved by expressing heat production per unit of surface area rather than per unit of weight. This is an example of the application of the so-called 'surface law', which holds that an animal's metabolic rate is proportional to its surface area and not to its weight. This is not true for all conditions (see Chapter 10), but it tends to hold for mature animals and it provides a useful basis on which to compare animals of different sizes. Thus in Fig. 5.5 approximate heat

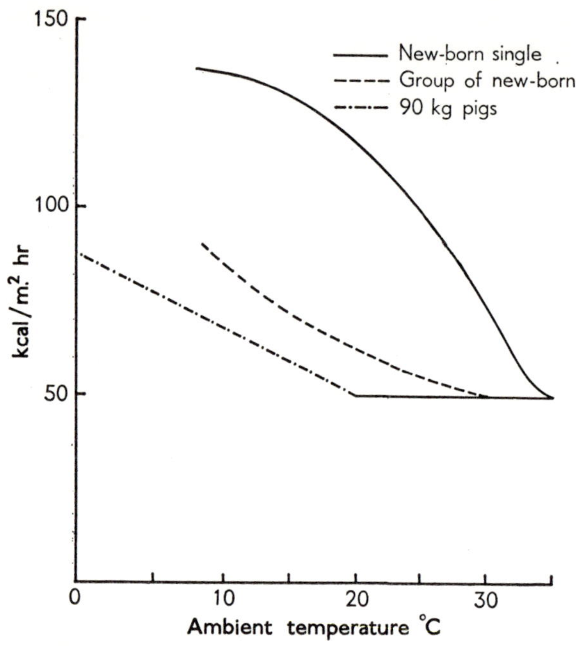

Fig. 5.5. The relation between metabolic rate and ambient temperature for single new-born pigs, a group of new-born pigs, and single 90 kg pigs (from Mount, 1965, by permission of European Association for Animal Production).

production rates per m² are given for single new-born pigs, a group of new-born pigs, and a 90 kg bacon pig. When metabolic rate is expressed on this common basis, the sharper reaction to cold of the new-born pig, compared with the older animal, becomes apparent; when the temperature falls the new-born reaches the upper limit of its metabolic rate earlier than the bacon pig.

The levels of minimal metabolism are similar, on a surface area basis, for new-born and 90 kg pigs; if instead they are compared on a weight basis (Table 5.2) the bacon pig's rate is about three times smaller than that of the new-born pig. As the pig grows in size, therefore, the total metabolism increases at a slower rate than the body mass. In order to obtain proportionality between metabolism and some function of body weight, the two quantities are often graphed on logarithmic axes (Chapter 10). By such means it is found that metabolism is most commonly proportional to body weight raised to a power less than unity; in the special case of proportionality to surface area, the power is 0·67. Table 5.2 also

TABLE 5.2

Rates of oxygen consumption expressed per kg body weight and per minute for pigs from birth to 100 kg in size. The rates given are the estimated resting maintenance energy requirements at thermal neutrality, so that they are minimal rates for the plane of nutrition (on which the rates depend—see Fig. 5.3) at the times of measurement. Values of columns *A* (fasting) and *B* (fed) derived from Brody (1945); *C* from Thorbek (1967); *D* from Mount (1959), Mount & Rowell (1960*b*) and Holmes (1966); respiratory quotients from Mount (1959) for new-born pigs, the other values calculated from Thorbek (1967)

Body weight kg	*Oxygen consumption ml./kg.min*				*Respiratory quotient Fed*
	A Fasting	*B* Fed	*C* Fed	*D* Fed	
1·2 (> 1 day old)				15·0	0·84
5		11·1		11·3	
10		10·1			
15	6·1	9·5			
20	5·7	9·4		8·7	
25	5·4	9·1			
35			10·7		1·06
50	4·7	8·5	9·2		1·09
60				6·1	
75	4·4	5·7	7·7		1·14
100	3·7	5·1			

gives values for the respiratory quotient in new-born and older pigs. In the rapidly growing pig, between 35 and 75 kg body weight, the respiratory quotient exceeds unity, which occurs in an animal which is producing fat from carbohydrate. In the new-born pig, the respiratory quotient is usually below unity.

Another consequence of metabolic rate being related more nearly to surface area than to body weight is that huddling of new-born pigs brings about a marked reduction in heat loss under cool conditions; the single bacon pig and the group of new-born pigs then become more closely comparable in terms of heat loss per unit area of each animal (Fig. 5.5). The effect of huddling on metabolic rate per unit body weight in pigs of 3 to 6 days of age is shown in Fig. 5.6. Oxygen consumption rate was measured at 20°C, and the number of pigs in the chamber was first progressively increased, and then decreased. There is an apparent inverse

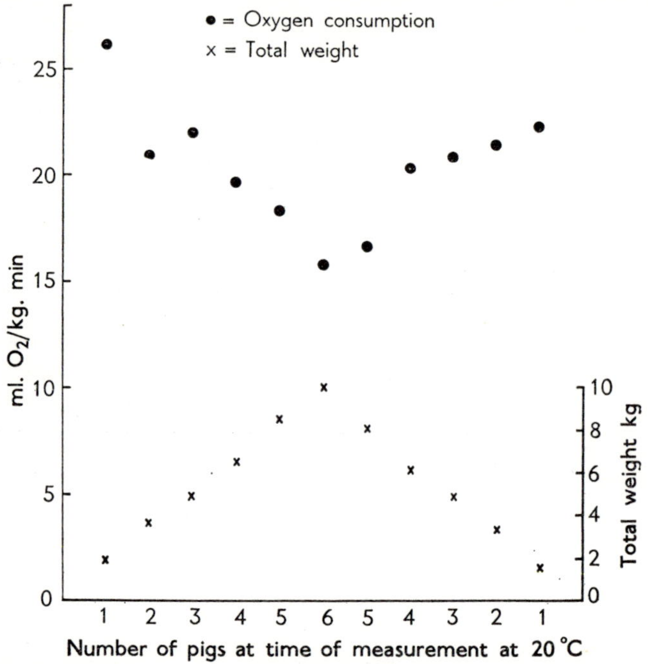

FIG. 5.6. The relation between metabolic rate per unit of body weight and the number of new-born pigs together in a chamber at an environmental temperature of 20°C (from Mount, 1960, by permission of *Journal of Agricultural Science*).

relation between the number of pigs together at the time of measurement and the level of metabolic rate. Under warm conditions, in thermal neutrality, pigs do not huddle together: Fig. 5.7 shows that a group of four or six pigs has a metabolic rate *per unit weight* at 35°C which is similar to that of a single pig alone in the chamber at that temperature. As the temperature falls, however, the pigs huddle together (see frontispiece) and the metabolic rate falls to levels characteristic of single larger pigs; the micro-climate of each animal is influenced considerably by the other pigs, and the thermal demand of the environment is correspondingly reduced. The result is that body temperature is maintained in the cold by metabolic rates of about 40% less than those which would be required for thermoregulation by the single pig in a similar chamber environment; the saving in energy expenditure becomes proportionately greater as the ambient temperature falls. Kleiber & Winchester (1933) found that one-to-three week-old chicks showed a reduction in metabolic rate of 15% as a result of huddling in groups of ten at 14 to 15°C.

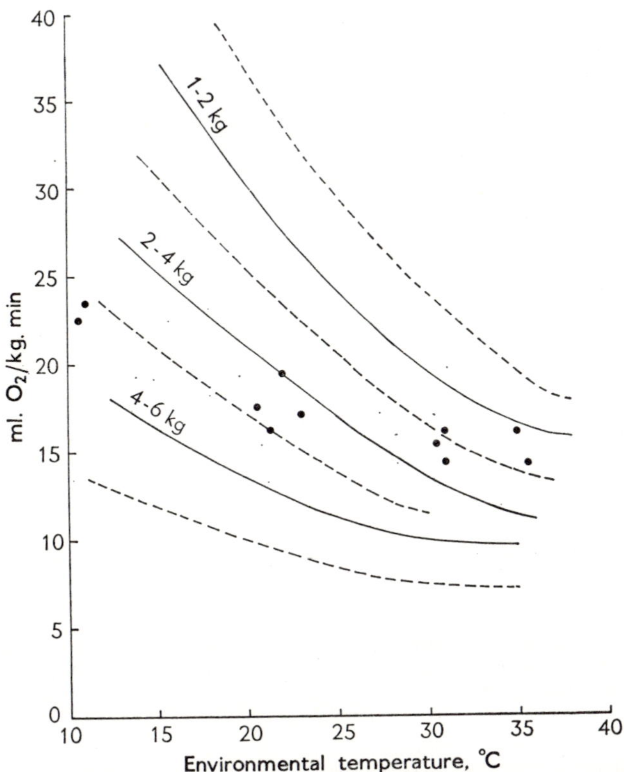

FIG. 5.7. Metabolic rates for three weight groups of young pigs over a range of environmental temperature. Curves refer to results on single pigs, with interrupted lines enclosing approximately 85% of results in each weight-class; points refer to results on groups consisting of either four or six animals, each weighing 1–2 kg, together in the metabolic chamber (from Mount, 1963a, by permission of Federation Proceedings).

HEAT LOSS FROM GROUPS OF GROWING PIGS

Huddling also leads to decreased heat losses in the cold when larger pigs are grouped together. The effect has been demonstrated during long-term measurements of heat loss in the Babraham pen calorimeter (Holmes & Mount, 1967).

The course of the experiments in the pen calorimeter allowed the comparison of three environmental temperatures, 9, 20, and

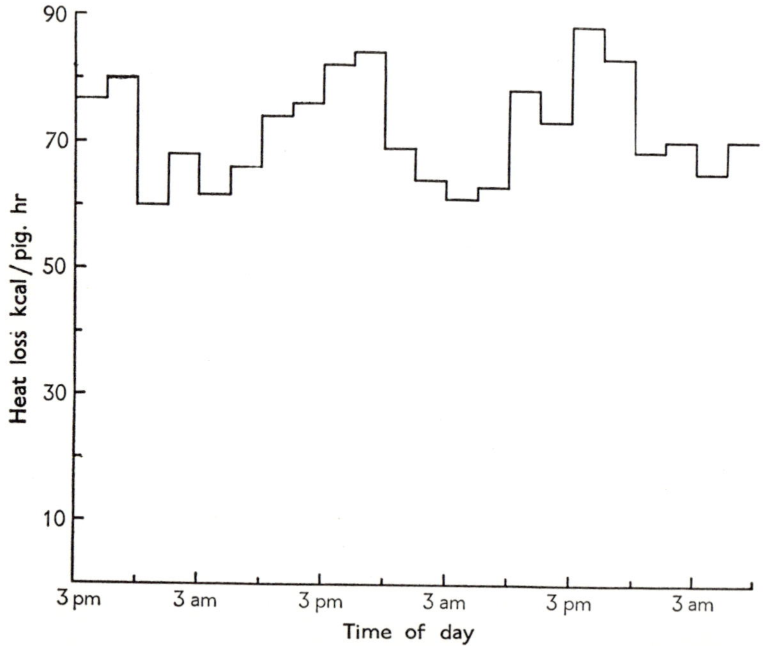

Fig. 5.8. 24-hour variation in heat loss from a group of six pigs, each weighing approximately 20 kg, at 20°C in the Babraham pen calorimeter (from Mount, 1967*b*, by permission of Lea and Febiger).

30°C, in their effects on heat losses from pigs at individual body weights of either 20 or 60 kg. Groups of three to six pigs were used in these experiments: food intake was at the level of 43 g/kg per day for the 20 kg pigs, and for the 60 kg pigs it was restricted in amount, in accordance with some farm practice, to 1·83 kg/pig per day. Fig. 5.8 shows the variation in heat loss from a group of pigs during several 24-hour cycles. The pig is a markedly diurnal mammal (except under very hot conditions), and although the animals were exposed to a constant intensity of light there were many other factors, such as noise and activity in the laboratory, to mark the 24-hour cycle. In addition, in the experiment shown the amplitude of the 24-hour heat loss cycle was probably increased by feeding taking place at 9.30 a.m. and 5.30 p.m. The group of pigs adapts to changes in environmental temperature more easily than

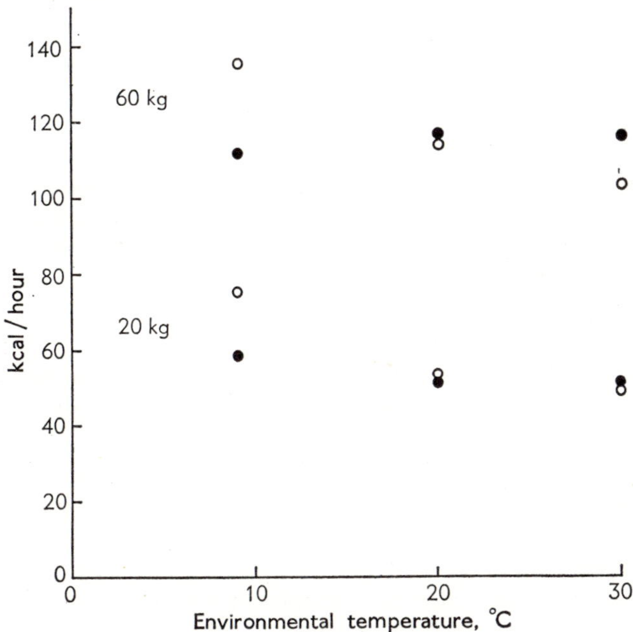

Fig. 5.9. Resting rates of heat loss from single pigs, ○, in a small calorimeter, and from groups, ● (3 pigs of 60 kg each, and 6 pigs of 20 kg each) of pigs in the Babraham pen calorimeter, at 9, 20, and 30 °C. The chief significant differences ($P < 0.01$), for both the single pigs and the groups, lie between 9 and 20 °C (Holmes, 1966).

the single pig, being able to break up or to coalesce in a huddle and so in this way to increase or decrease heat loss. Fig. 5.9 summarizes results on rates of loss of heat both from groups and from single pigs at 9, 20, and 30 °C. Heat loss at 9 °C is increased for pigs of both 20 and 60 kg body weight, and more in the case of the single pigs than in the groups (Holmes, 1968). At 20 and 30 °C, the group heat losses are similar, although inspection shows the pigs to be lying in a group at 20°C, and to be spread out at 30°C.

24-hour cycle

Cairnie & Pullar (1959) made an investigation into 24-hour periodicity in metabolic rate in young piglets in a gradient layer calorimeter. In some experiments they attempted to maintain a constant environment during the 24 hours of each day by leaving a light on continuously, making food and water available at all times, and by keeping the animals and the calorimeter in a partially sound-proofed room, although the pigs were disturbed more frequently during working hours than at night. The departures of heat output from the mean 24-hour values in these experiments were compared with others in which pigs were subjected to 24-hour periodicity of exposure to light (Fig. 5.10); both groups were at constant temperature. It is clear from the figure that the 24-hour cycle was not abolished in the first group; in the second group the cycle was more pronounced.

Bond, Kelly, & Heitman (1967) exposed pigs to constant temperatures or to temperatures varying within a 24-hour cycle; the animals were fed morning and evening. They found the ranges of rectal and skin temperatures, and respiration rates, to vary proportionately to the range of environmental temperature variation. Even in a constant environment, however, the ranges of rectal temperature and respiration rate were still considerable during the 24 hours—about 0·3°C and 15 respirations per minute, respectively.

The considerable 24-hour variation in heat loss and activity in groups of pigs might also be influenced by 24-hour variations in environmental temperature. Sørensen (1962) found that in its effects on weight gain for a given food intake, rhythmic temperature variation corresponded to a constant temperature close to the mean value of the fluctuating temperature during the experiment. His results also show the greater weight gains achieved with pigs grouped together, and having straw available to them, as compared with pigs individually housed (see Table 5.5). Bond, Kelly, & Heitman (1963) found that constant levels of 21 and 32°C were more conducive to efficient and rapid animal weight gains than cycling temperatures of the same average levels. There was, however, no significant difference in heat and water losses between cycling temperatures and a constant mean temperature. The effects of sine wave temperature variations on rectal and surface

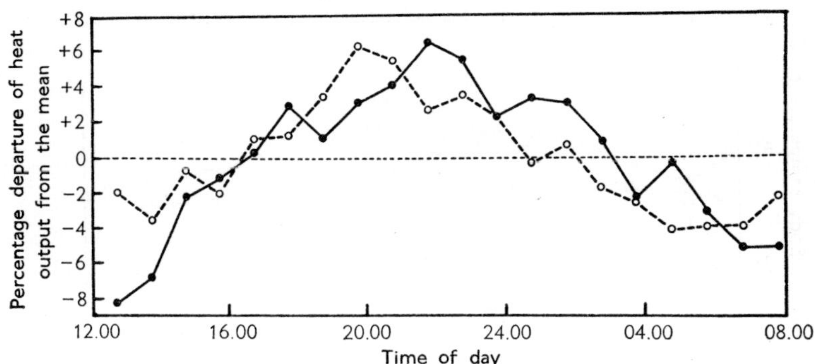

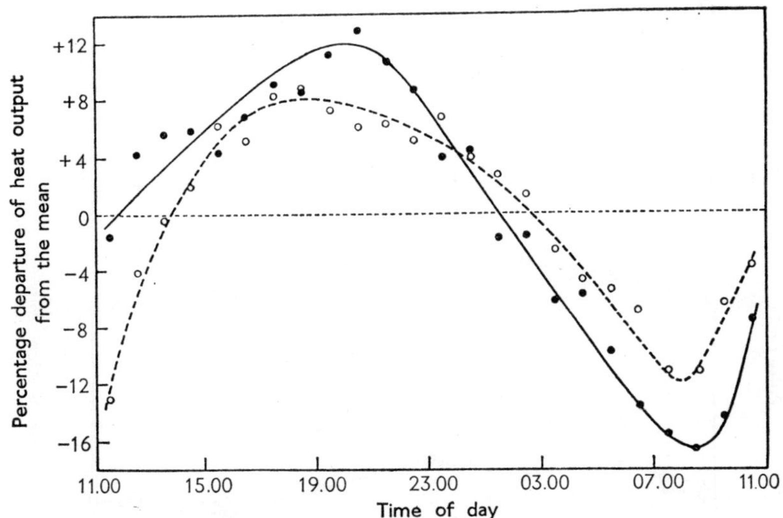

FIG. 5.10. Upper diagram: mean cycle of heat output for three pigs subjected to a régime without 24-hr periodicity.

○—○ : weight range 3 to 6·5 kg;
●—● : weight range 9 to 11·5 kg.

Lower diagram: mean cycle for eight pigs subjected to a régime with 24-hr periodicity. Weight range 3 to 11·5 kg.

○—○ : runs beginning at 11·00 hrs;
●—● : runs beginning at 23·00 hrs

(from Cairnie & Pullar, 1959, by permission of *British Journal of Nutrition*).

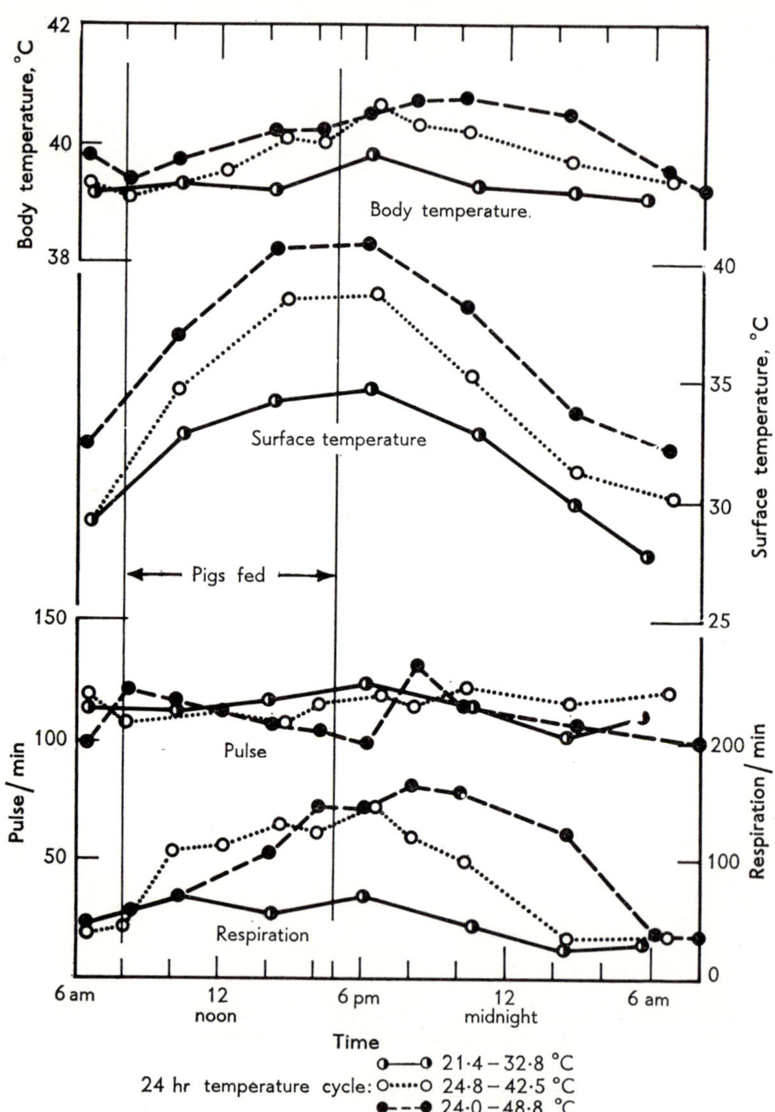

FIG. 5.11. The variations in body temperature, surface temperature, pulse rate, and respiration rate in pigs exposed to 24-hr sinusoidal temperature cycles of three amplitudes and mean values (from Bond, Kelly & Heitman, 1967, by permission of *Animal Production*).

temperatures and pulse and respiration rates are shown in Fig. 5.11.

The work of Bond *et al.* (1963) in California differs from that described by Sørensen (1962) in two important respects. First, the

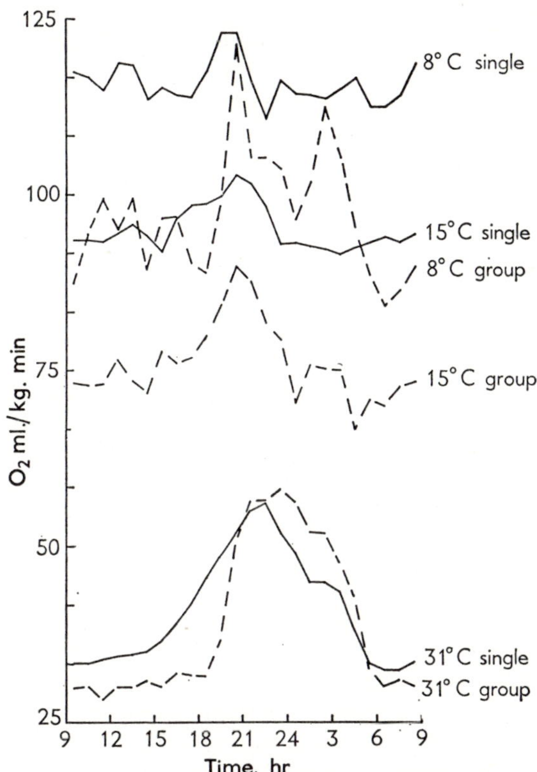

FIG. 5.12. Mean oxygen consumption rates over the 24 hr period for both single mice and groups at environmental temperatures of 8, 15, and 31 °C. Single mice: continuous lines; groups of mice: interrupted lines. Note that ordinate scale does not extend to zero (from Mount & Willmott, 1967, by permission of *Journal of Physiology*).

Californian pigs were fed all they could eat on two occasions each day, whereas in Sørensen's work the food intake was controlled. Second, Sørensen's environmental temperatures tended to be lower. It is therefore not possible to compare the two sets of results, although both suggest that a limited and regular 24-hour variation in environmental conditions would not have a very marked effect on the mean energy exchange.

An investigation into the relation between spontaneous activity, metabolic rate, and the 24-hour cycle, in mice at different environmental temperatures, revealed marked variation in both metabolic rate and activity (Mount & Willmott, 1967). The maxima occurred during the night, in keeping with the animal's nocturnal habit. Oxygen consumption rate in the cold was diminished for each mouse in a group compared with mice kept singly, but was unchanged by grouping at the critical temperature of 31°C (Fig. 5.12). The amplitude of the 24-hour variations decreased at the lower environmental temperatures of 8 and 15°C, when the high thermal demand of the environment required high metabolic rates even when the animals were resting grouped together.

SKIN TEMPERATURE AND METABOLIC RATE

Hart (1963) has shown a very large species variation in the level of surface temperature at which heat production is stimulated (Fig. 5.13); furred animals are metabolically sensitive to lowering of skin temperature. Within a given species, however, the correlation between surface cooling and thermogenesis may be quite precise.

Skin temperature operates in two different ways, one as a factor in thermoregulation through its effect on peripheral receptors, and the other as a factor in determining heat exchange with the environment. In the pig, skin temperature is closely related to environmental temperature in consequence of the animal's lack of coat, as shown in Fig. 5.14 (and see Fig. 4.8). The relation between local heat loss from the body surface measured by a heat-flow disc and the temperature gradient between the body and the environment is shown in Fig. 5.15, and here the effect of the rise in skin temperature in slowing down the fall in heat loss rate at the higher environmental temperatures is shown in the second part of the diagram. At low temperatures the phenomenon of cold-induced vasodilatation occurs in the pig (Fig. 5.16). This phenomenon is well known in man (Burton & Edholm, 1955), and has been seen, in the form of ear temperature fluctuations, in the adult pig by Irving (1956*a*) and in the calf by Ingram & Whittow (1962). There is considerable variation between pigs in the degree to which ear temperature varies in the cold.

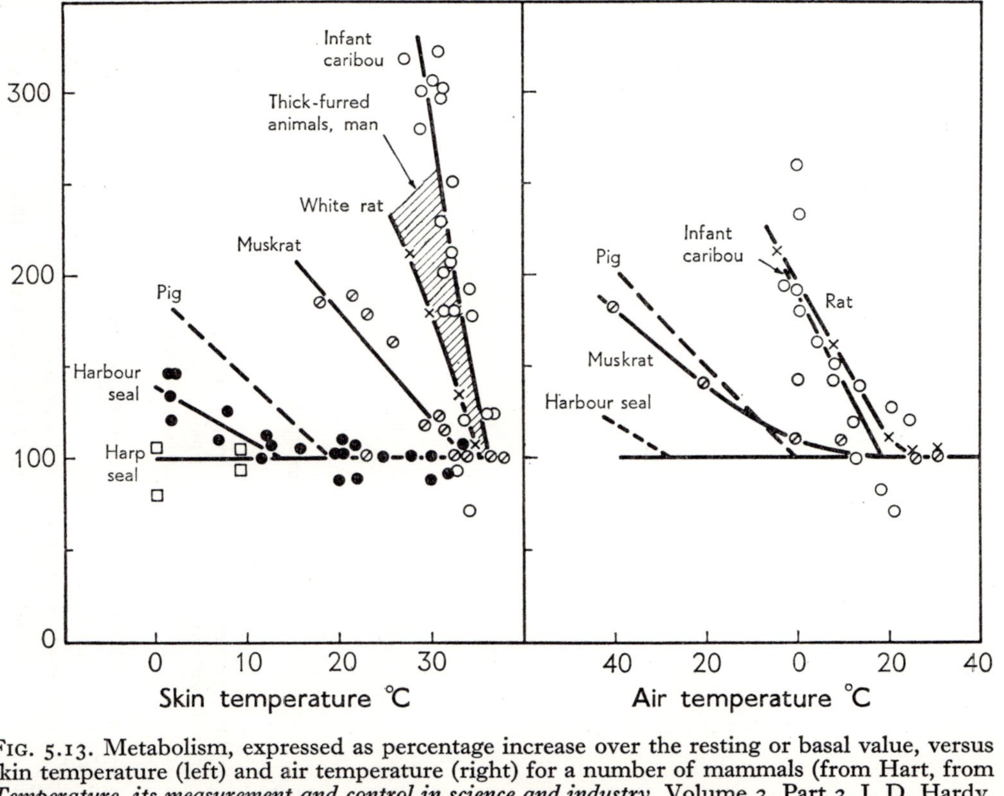

FIG. 5.13. Metabolism, expressed as percentage increase over the resting or basal value, versus skin temperature (left) and air temperature (right) for a number of mammals (from Hart, from *Temperature, its measurement and control in science and industry*, Volume 3, Part 3, J. D. Hardy, Editor, Reinhold Publishing Corporation, New York, 1963).

Fig. 5.17 shows the relation of skin temperature to air temperature for large pigs compared with man and furred animals. The diagram shows the differences in skin temperature, at given environmental temperatures, between Alaskan and Californian

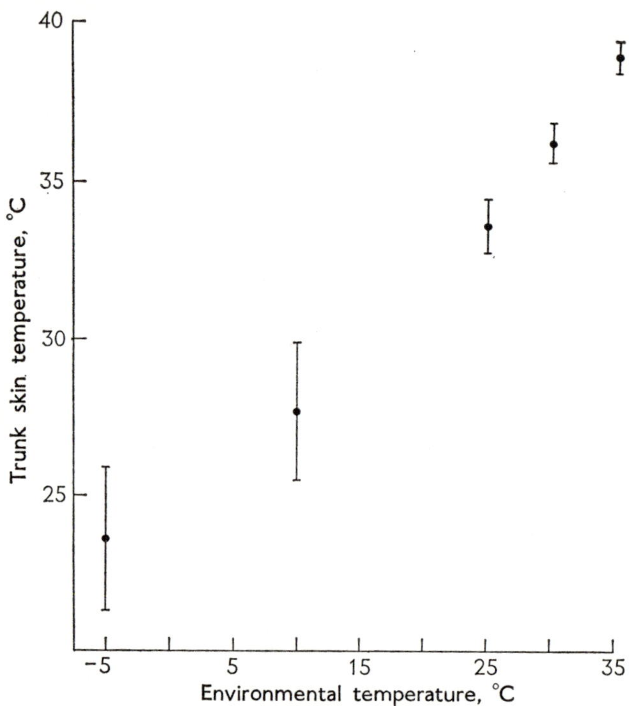

FIG. 5.14. The relation between environmental temperature and skin temperature on the trunk of the pig (mean ± standard deviation) (from Ingram, 1964a, by permission of *Research in Veterinary Science*).

pigs, and the contrast between the pig and a thickly furred animal. Values for naked and clothed man are included for comparison.

The heat increment of feeding is associated with a significant rise in skin temperature, the degree and duration of this appearing to depend on the amount of food taken (Fig. 5.18). Neergaard & Thorbek (1967) measured at twelve-minute intervals the metabolic rate of growing pigs, at three body weights, for a period of two hours following feeding. The rate of heat production reached a maximum at about three-quarters of an hour after feeding, and

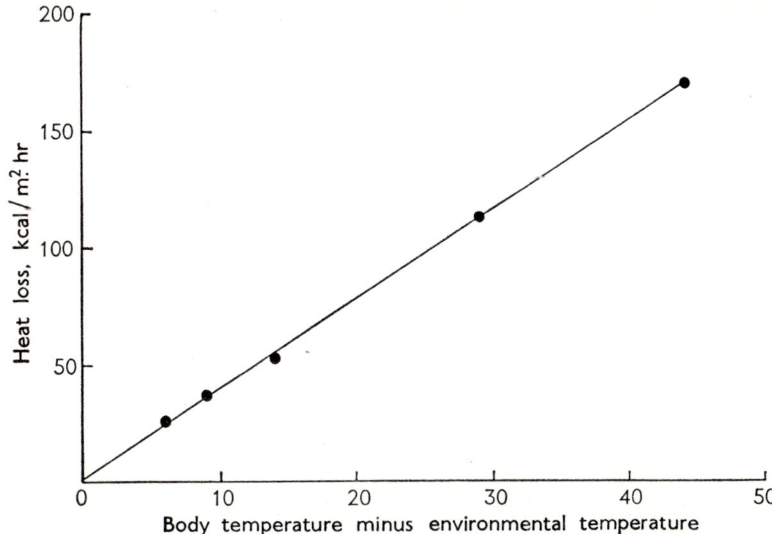

FIG. 5.15. (*a*) The relation between heat loss from the pig and the temperature gradient between the body and the environment. (*b*) The relation between heat loss and environmental temperature (from Ingram, 1964*b*, by permission of *Research in Veterinary Science*).

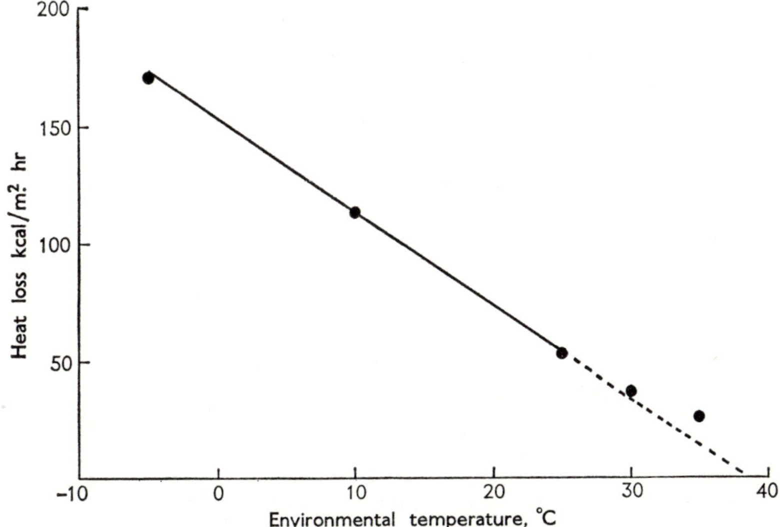

FIG. 5.15. (*b*)

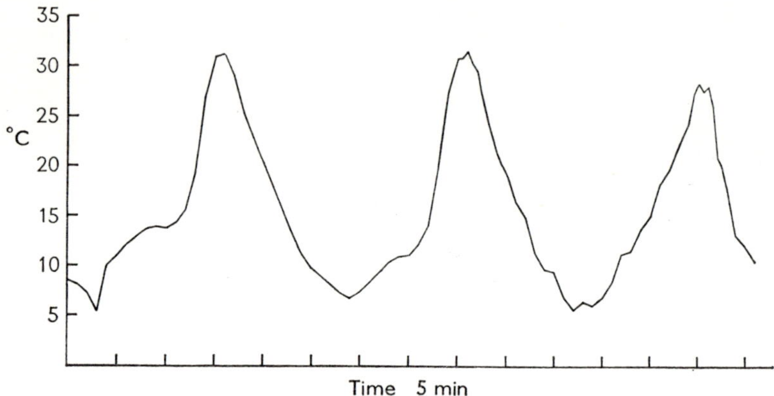

FIG. 5.16. The variations in skin temperature on the pinna of the ear of the pig at an environmental temperature of −5°C (from Ingram, 1964*a*, by permission of *Research in Veterinary Science*).

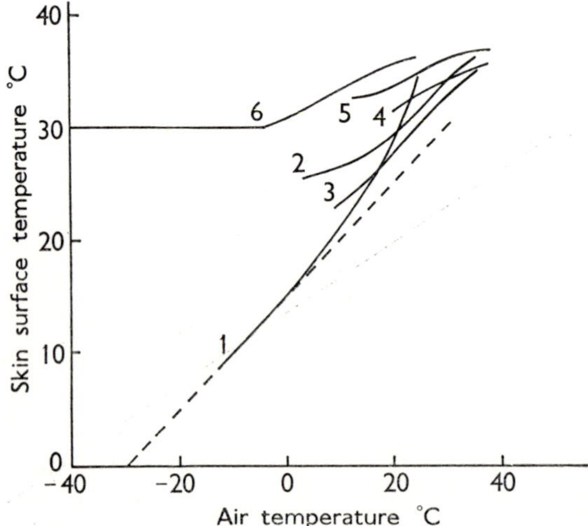

FIG. 5.17. Curves relating skin and air temperature in several mammals. 1, Alaskan pigs; 2, Californian pigs; 3, naked man working; 4, naked man resting; 5, clothed man; 6, thick furred arctic mammals (from Irving, 1956*a*, by permission of *Journal of Applied Physiology*).

then declined slowly (Fig. 5.19). The two diagrams, from different laboratories, reflect the marked effect of feeding on heat loss from the pig.

The thermal circulation index has been used to determine the critical temperature in the older pig in a manner similar to its use

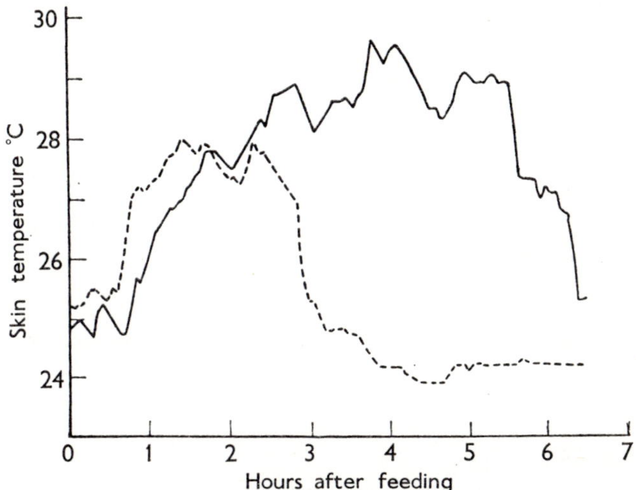

FIG. 5.18. Variations in skin temperature following food intake in pigs of 26 kg body weight at air temperature of 3°C. Continuous line: intake of o·5 kg barley and 100 g milk powder; broken line: intake of o·25 kg barley and 50 g milk powder (from Sørensen, 1962, by permission of Royal Veterinary Agricultural College, Copenhagen).

in the new-born pig (see Fig. 4.9). Ingram (1964*b*) found the tissue insulation to change considerably between environmental temperatures of 25 and 30°C, and in this respect to parallel the change in the thermal circulation index (see Table 5.3). The pulse-rate in

TABLE 5.3

Mean values of thermal circulation index and tissue insulation from measurements on seven 3-month-old Landrace pigs, over a range of environmental temperature. Tissue insulation calculated from ratio of rectal–skin temperature difference to heat loss per unit area (from Ingram, 1964*b*)

Environmental temperature, °C	−5	10	25	30	35
Thermal circulation index	1·83	1·57	1·58	2·32	2·44
Tissue insulation, °C.m².hr/kcal	o·101	o·107	o·113	o·085	o·063

3-month-old pigs increases from about 80/min at temperatures below 25 to 30°C to values exceeding 100/min at 35°C (Ingram, 1964*a*).

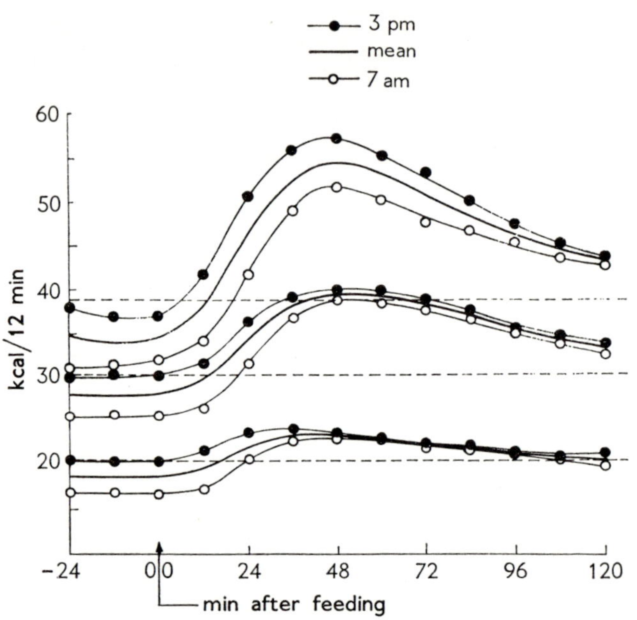

FIG. 5.19. Heat production of growing pigs fed at 3 p.m. and 7 a.m.
compared with the mean 24-hour heat production rate (— — —),
at 18–20°C. Mean body weights of pigs, with reference to the
curves, from above downwards: 85, 54, and 30 kg (from Neergaard
& Thorbek, 1967, by permission of European Association for
Animal Production).

TEMPERATURE ADAPTATION

Prosser (1964) uses the term 'physiological adaptation' to mean
any property of an organism which favours survival in a specific
environment, particularly a stressful one. Both physiological and
behavioural reactions, as these are commonly understood, play a
part in the adaptation of the pig to variations in its thermal environment. The interest attached to thermal adaptation in this species lies
not only in the ways in which survival may be favoured; considerable
interest is also derived, for economic reasons associated with pig
farming, from the effects which such adaptation might have in
influencing the utilization of food in growth as opposed to its use
for maintenance purposes.

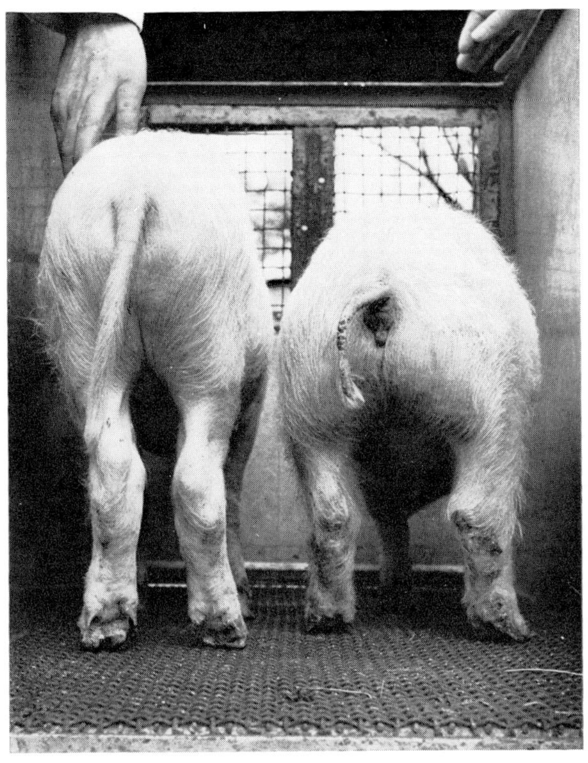

FIG. 5.20. Two litter-mates, the pig on the left exposed to 35 °C for seven weeks, and the pig on the right to 5 °C; the pig in the cold was fed an amount to keep it at approximately the same body weight as the pig in the warm (by courtesy of Dr D. L. Ingram & Dr M. E. Weaver).

Acute responses to the thermal environment, such as the rise in metabolic rate on exposure to cold, constitute short-term adaptational changes. In addition to these, longer-term exposures to given environments produce more long-lasting reactions, including morphological as well as metabolic responses (Barnett & Mount, 1967). A discussion of some aspects of the long-term adaptation of the pig to its thermal environment follows.

Adaptation to heat

Acclimatization to hot environments can take place in man and animals as a result of changes in sweating and in the circulation. The pig, however, does not sweat effectively under hot conditions, so that this mode of adaptation is denied to it. Part of the acclimatization of cattle to heat appears to involve a reduction in food intake, and it is well known that the food consumption of pigs falls sharply in hot surroundings. There is the possibility that the pig's metabolic rate falls during prolonged exposure to heat, and the question which arises therefore is whether such a reduction in rate takes place, and, if so, to what extent it is related to the decreased food intake under hot conditions.

In a series of experiments (Ingram & Mount, 1965) pigs aged 12 days were weaned on to a pelleted diet, and 6 days later were exposed individually in cages to environmental temperatures of 17, 25, 33, and 35°C for a further two weeks. Litter-mates were divided between experiments run in pairs of temperatures: 17 and 27°C, 27 and 33°C, and 25 and 35°C. In some cases the animals were allowed to feed *ad libitum*, while in others the food intake in the cooler group was restricted to that taken voluntarily by pigs in the warmer group.

Oxygen consumption and body temperature observed from measurements made over a range of ambient temperature (6 to 35°C) at the end of the two weeks' exposure indicated that in the case of each pair some degree of adaptation to the hotter environment had occurred. This was shown by the finding that sojourn at the higher of the two temperatures was followed by a lower oxygen consumption rate when this was measured for each pair at a given common temperature. However, control of food intake removed the difference in metabolic rate between those pigs kept at 17 and 27°C, and between those kept at 27 and 33°C. In the 25 and 35°C experiment, there was still a difference in metabolic rate even when food intake was controlled. The level of food intake therefore

E

contributes considerably to apparent heat adaptation, although there is evidently a small independent element of metabolic acclimatization.

One possibility is that the reduction in metabolism of animals living above their critical temperature is related in part to a diminished rate of secretion of thyroid hormone. A comparison was made between pigs reared at 25 and 35°C with respect both to the secretory activity of the thyroid gland and to oxygen consumption (Ingram & Slebodzinski, 1965). Thyroid secretory activity was estimated from the rate of release of ^{131}I from the gland and by the conversion ratio of inorganic to organic iodine. Secretory activity was always greater in the animals housed at 25°C, and the oxygen consumption rate of these animals was also greater than the rate of the pigs kept at 35°C.

When the thyroid gland was removed at 8 days of age, and a small dose of 1-thyroxine ($10\mu g/kg$) given each day, there were no differences in metabolic rate between pigs housed at 25 and 35°C, whether the animals were fed *ad libitum* or had their food intake controlled. In the absence of the thyroid gland, therefore, metabolic adaptation is impaired. Under hot conditions such adaptation, when separated from the effects of variation in food intake, is in any case of a small order.

Adaptation to cold

There is little definitive work on adaptation to cold in either pigs or other farm animals. Much more work of this sort has been done on laboratory animals, although farm animals in many parts of the world live at winter temperatures well below thermal neutrality (Hess & Bailey, 1961).

Irving (1956*a, b*), who investigated the skin temperature and thermal insulation level of pigs kept in the Arctic, demonstrated the ability of the animal to develop considerable cold-resistance in spite of the lack of a coat. In discussing the adaptation of pigs to cold, Irving (1964) points out that the animals can live out of doors with only the protection of a roof and dry bedding. In pigs adapted to the cold, the skin over the entire body may cool below 5°C in air at −30°C, but the cold skin does not appear to cause an increased metabolic rate, and the animals seem to be comfortable. In large pigs, the temperature gradient fom the cold skin to the core temperature may occupy up to 10 cm of tissue; this distance, and the magnitude of the gradient, vary with the temperature of the air (see Fig. 3.3). Irving found the critical temperature of 50 kg boars in summer to be 0°C. The temperature of the skin shows periodic warming and cooling, as mentioned earlier, similar to the periodic vasodilatation which may occur in cold fingers (Lewis, 1930).

Sørensen's (1962) results suggest that in the cold, pigs may lay down subcutaneous fat as an adaptation, even when the maintenance requirements of a cold environment would not normally permit fat storage. Fuller (1965) found that young pigs kept at 10°C had significantly more hair than those in warmer environments, and that the animals at 30°C had larger ears than pigs kept at lower temperatures.

The general effects of exposing pigs to hot and cold environments are illustrated in Fig. 5.20 (D. L. Ingram & M. E. Weaver, personal communication). The two animals, litter-mates, were weaned at 12 days of age, and put into controlled-temperature rooms at 20°C. After two weeks, the temperature of one room was dropped to 5°C, and in the other was raised to 35°C. During the following seven weeks the 5°C pig was allowed food *ad libitum* and the 35°C pig was fed an amount to keep it at approximately the same body weight as the pig in the cold. Among the differences in appearance, the shorter limbs and apparently denser coat of the 5°C pig are outstanding; the hair density was 8·9 mg/cm^2 in the 5°C pig, and 6·8 mg/cm^2 in the 35°C pig. Two other pigs treated similarly showed similar effects.

Slee & Sykes (1967) found good evidence for cold acclimatization in sheep. They exposed closely-shorn sheep, either on high or low planes of nutrition, to short acute cold exposures, down to -20°C, separated by a period of two weeks in either a thermoneutral ($+30$°C) or a cool ($+8$°C) environment. 37 out of 40 sheep showed a greater resistance to body cooling at the second exposure; the mean rates of fall of rectal temperature (°C per 100 min exposure) were 0·42 at the first exposure and 0·22 at the second exposure (Fig. 5.21). There was great individual variation in initial cold resistance and in ability to acclimatize.

GROWTH AND METABOLISM

The required protein content and other components of diets for pigs have been reviewed by others (Lucas, 1964; Cunha, 1966; Agricultural Research Council, 1967). Pond, Barnes, Reid, Krook,

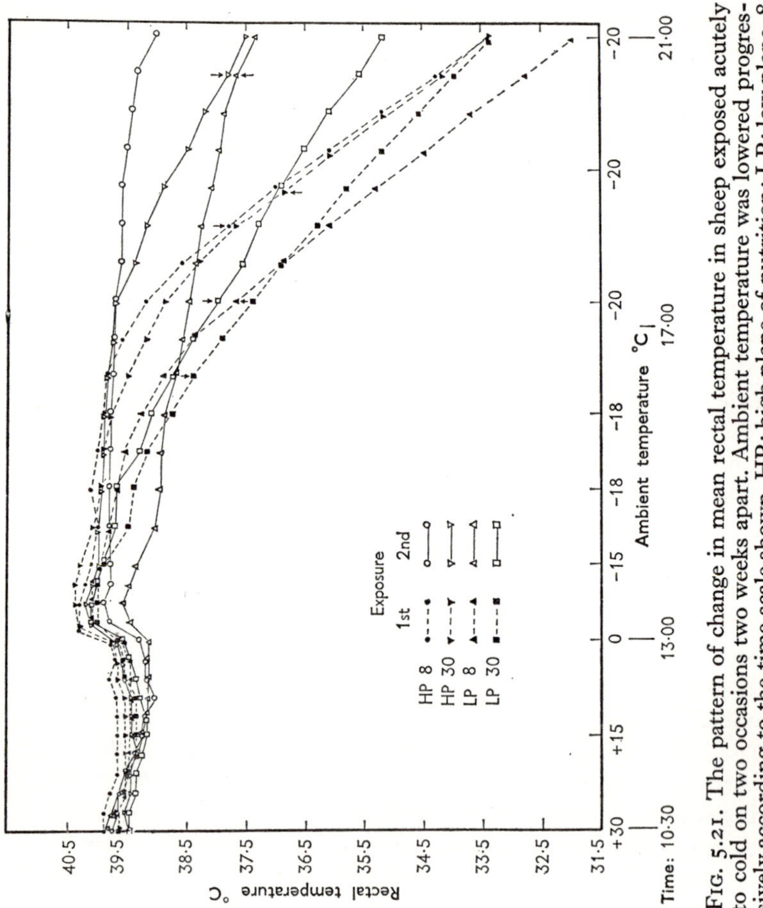

Fig. 5.21. The pattern of change in mean rectal temperature in sheep exposed acutely to cold on two occasions two weeks apart. Ambient temperature was lowered progressively according to the time scale shown. HP: high plane of nutrition; LP: low plane. 8 and 30 refer to the temperatures, °C, at which the sheep were kept between the cold exposures (from Slee & Sykes, 1967, by permission of *Animal Production*).

Kwong & Moore (1966) weaned pigs at three weeks of age (3 to 7 kg body weight) on to a 3% protein diet. This was found to produce a kwashiorkor-like syndrome, with anaemia, gross oedema, reduced serum protein, and apathy. When the pigs on a low-protein diet were fed a low level of fat they tended to remain active and alert. The combination of low-protein and high-fat in the diet results in greater fatty infiltration of the liver, more severe depression in serum proteins and greater oedema and apathy.

Young pigs can be kept at body weights of about 6 kg for long periods, up to a year, simply by feeding them too little of a diet which would otherwise be satisfactory (McCance, 1960). The resulting undernourished animals produce less heat than normal animals of the same size, and have lower deep body and skin temperatures. When exposed to an environmental temperature of 11°C, they can increase their heat production considerably above their heat production at their housing temperature of 24°C (McCance & Mount, 1960). During the first days of rehabilitation on a normal level of food intake, the rectal and skin temperatures rise and approach those of normal pigs, body weight increases rapidly, and the rate of oxygen consumption increases (Mount, Lister & McCance, 1963).

The effects of high and low temperatures on metabolic rates have their counterparts in the changes seen in growth rate and body composition. Sørensen (1962) refers to experiments in which groups of pigs were kept on given intakes of food at different environmental temperatures; daily weight gain was estimated, and the rate of protein deposition determined by estimation of nitrogen balance. Protein deposition was highest at 15–23°C, and was considerably reduced at lower temperatures. Weight gain was also less in the cold. Another result of exposure to cold is a higher thyroid secretion rate (Table 5.4), and a higher food conversion ratio (that is, ratio of food eaten to weight increment).

Weight gain and thyroid secretion rate at a given feeding level are modified considerably by grouping pigs together and by providing straw bedding (Table 5.5). In this connection, Jones, Pickett, Kadlec & Morris (1965) found that pigs with bedding gained weight at a rate of about 0·1 kg/day faster than pigs without bedding, in the course of a winter experiment. The pigs in the bedded area also consumed 2 kg less feed per 10 kg weight gain, so that their food conversion efficiency was enhanced. The effects of grouping pigs and providing bedding can be referred, in part at

TABLE 5.4

The influence of climate on thyroid secretion rate and on food conversion ratio in pigs (from Sørensen, 1962)

Climate		Thyroid secretion rate*		Scandinavian feed units per kg gain from 40 to 90 kg body-weight	
Temperature °C	% relative humidity	mg thyroxine per day	Number of pigs	Feed units per kg gain	Number of pigs
24	90	0·38 ±0·03	12	3·5 ±0·07	24
23	50	0·43 ±0·03	5	3·4 ±0·08	15
15	70	0·51 ±0·08	5	3·4 ±0·05	16
8	70	1·72 ±0·18	12	3·8 ±0·09	16
3	70	3·81 ±0·67	5	4·3 ±0·20	5

* Corrected to 100 kg body-weight by correction linear with (body-weight)[0·7].

TABLE 5.5

The influence of straw bedding and of huddling upon daily gain in weight, food conversion ratio (40–90 kg body-weight) and thyroid secretion rate in pigs at an ambient temperature of 3°C (from Sørensen, 1962)

	Daily gain in weight kg	Scandinavian feed units per kg weight gain	Thyroid secretion rate mg thyroxine per day*	Number of pigs
Individual pens, no straw	0·45 ±0·005	5·9 ±0·03	6·34 ±0·66	3
Group, no straw	0·63 ±0·034	4·3 ±0·20	3·81 ±0·67	5
Group, straw	0·72 ±0·014	3·7 ±0·05	2·41 ±0·15	5

* Calculated for 100 kg body-weight by linear correction with (body-weight)[0·7].

least, to the establishment of a warmer micro-climate with a smaller environmental thermal demand, so that less energy is dissipated in thermoregulation. In the Danish experiments (Sørensen, 1962) it was found that at low temperatures the amount of protein deposition was less relative to the daily weight gain. In agreement with this finding, chemical analysis of entire pig carcasses revealed that in those pigs raised at a lower temperature the fat content was higher.

Exposure of pigs to a sudden change in environmental temperature has resulted in marked changes in nitrogen excretion (Fig. 5.22), weight gain and thyroid secretion. Sørensen concludes from this observation that pigs require a period of three to four weeks to be acclimatized to new climatic conditions.

Slebodzinski (1965c) determined the thyroxine utilization rate in pigs from 10 to 34 days of age, and found the rate, when expressed per kilogram, to decrease with advancing age. Metabolic rate per kilogram also decreases with increasing body size. The daily

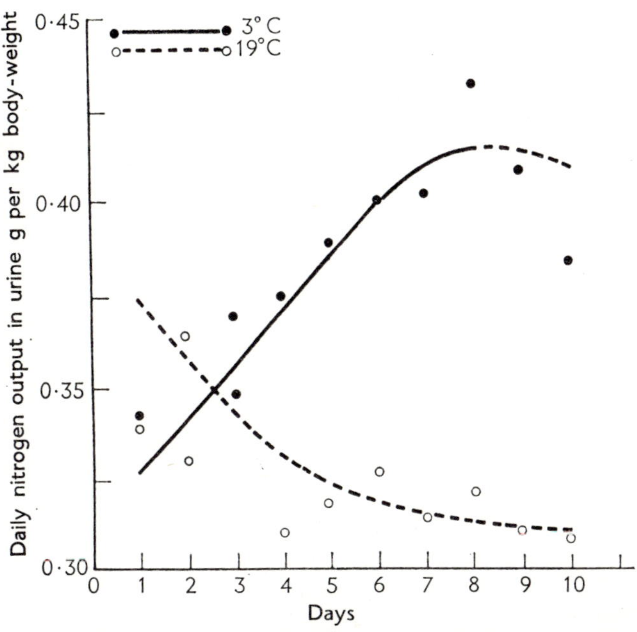

FIG. 5.22. Daily nitrogen output in urine in pigs after abrupt changes of air temperature from 19 to 3°C (continuous line) and from 3 to 19°C (broken line). Each dot represents an average from 12 experiments (from Sørensen, 1962, by permission of Royal Veterinary Agricultural College, Copenhagen).

utilization of thyroxine was 28·6 ($\pm$2·2) μg of thyroxine per kg in 10-day old piglets, and 21·2 ($\pm$2·3) μg/kg in 34-day old pigs. Figure 5.23 shows the thyroxine utilization rate for animals of different age and body weight.

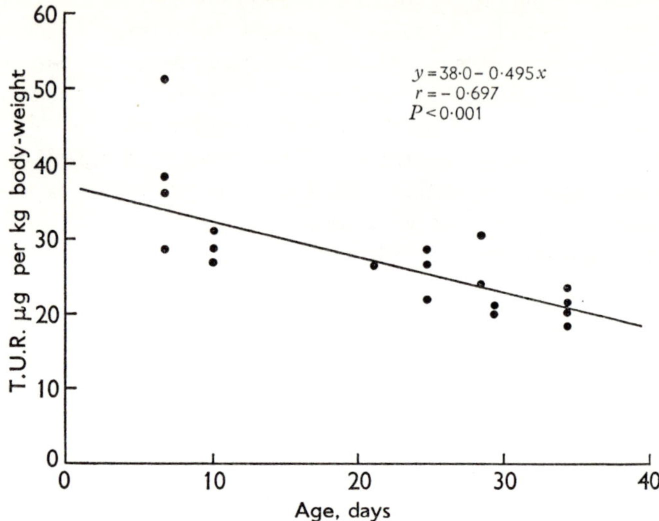

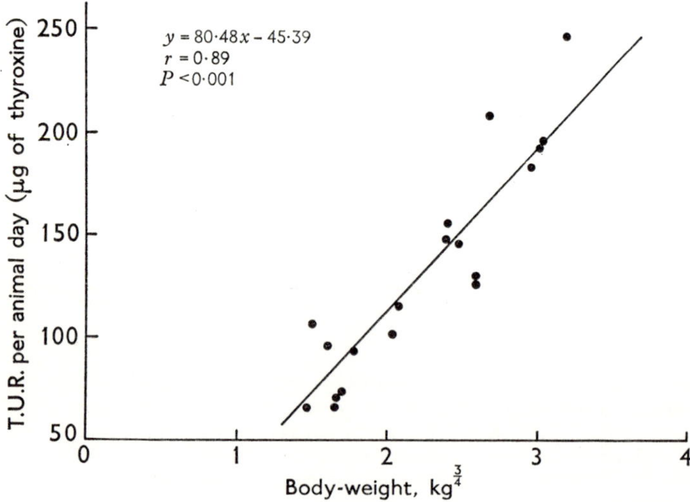

FIG. 5.23. The relations between thyroxine utilization rate (T.U.R.) and age and body weight in 19 Large White pigs 7 to 34 days old. Upper diagram: T.U.R. per kg against age; lower diagram: T.U.R. per animal against weight to the three-quarters power (from Slebodzinski, 1965c, by permission of *Research in Veterinary Science*).

BODY COMPOSITION

For economic reasons a considerable volume of work has been done on the body composition of pigs. In view of the pig's importance as a meat-producing animal, many of the measurements have been of the carcass analysis type, in which bone, muscle, and fat are separated out by dissection and weighed. Another approach has been the chemical analysis of the whole carcass or of samples of the minced carcass. These are direct methods, which are reliable but which suffer from the disadvantage that because the animal must be killed and the carcass destroyed, it is not possible to study the time-course of changes in body composition in one animal; it is instead necessary to use large numbers of animals compared statistically.

The changes which occur in body composition in the pig are particularly rapid in the period following birth (Table 5.6). The

TABLE 5.6

Changes in the body composition of the pig in the period following birth. Protein, lipid, water, and ash are given as percentages of the weight of the skinned carcass (from Brooks, Fontenot, Vipperman, Thomas & Graham, 1964)

Age days	Body weight kg	Protein %	Lipid %	Water %	Ash %
Birth	1·1	11·7	1·4	74·5	5·1
6	2·0	14·2	9·7	71·1	3·5
12	3·4	15·0	15·3	66·2	3·3
18	3·8	15·5	13·7	66·2	3·2
24	8·0	15·0	17·3	63·5	3·2
30	12·1	13·8	18·4	63·8	3·1

most outstanding feature is the rapid rise in the percentage of fat in the animal, from about 1% at birth (Widdowson, 1950) to about 10% at one week of age. The fat content of the carcass continues to rise steadily after that time, but is to some degree dependent on the nature of the diet (Fig. 5.24). The percentage of water in the body shows a corresponding decline during the first weeks following birth.

The intensive selection of pigs as meat-producing animals, and their maintenance on restricted levels of food intake, has led to the production of animals which over the decades have tended to contain less fat. None the less, the rates of deposition of fat are in general higher than those for protein on a body-weight basis, except for the few weeks following birth (Table 5.7). Figures for the physical composition of pigs at different weights (Cuthbertson & Pomeroy, 1962; Brooks, Thomas, Kelley, Graham & Allen,

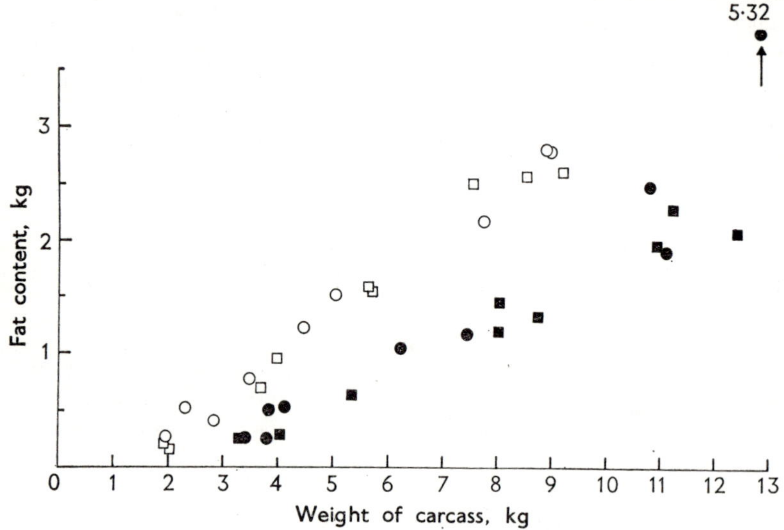

FIG. 5.24. Fat content of pig carcasses as a function of carcass weight, sex, and diet during the first six weeks following birth. 14% protein diet, males □; females ○. 50% protein diet, males ■; females ● (from Filer, Owen & Fomon, 1966, by permission of Frayn Printing Company, Seattle).

1964) suggest that muscle production occurs at a decreasing rate following the attainment of 70 kg body weight, but that the body content of fat does not exceed that of muscle before 120 kg body weight is reached. Fat tissue, however, contains little water, whereas muscle contains a great deal, and for energy balance purposes it is the protein content, and not the whole muscle, which must be considered.

In addition to the direct determination of body composition, a number of indirect methods have been developed. These methods allow animals to be followed in the course of growth, but on the whole their reliability in predicting body composition is low. Specific gravity measurements have given correlations of -0.57,

−0·58 and −0·30 with body fat by chemical analysis at 27, 55, and 90 kg body weight in live pigs; the same procedure, applied to the eviscerated carcass, gave a higher correlation of −0·75 (Kay & Jones, 1964). The packed red cell volume has been used as an index of 'lean body mass' in pigs, yielding a correlation of 0·82 (Doornenbal, Asdell & Wellington, 1962). The determination of

TABLE 5.7

Relative protein and fat deposition in pigs;
mean of seven pigs for protein, three pigs for fat
(from Oslage & Fliegel, 1965)

Body weight kg	Protein deposition g/kg.day	Lipid deposition g/kg.day
20–30	4·6	—
30–40	3·1	3·4
40–50	2·3	4·2
50–60	2·0	4·7
60–70	1·8	5·1
70–80	1·5	5·3
80–90	1·3	4·5
90–100	1·1	4·0
100–110	1·0	3·9
110–120	0·9	3·5
120–130	0·9	3·2
130–140	0·7	2·7
140–150	0·7	2·5
150–160	0·6	2·2

^{40}K and body water have also been used as indices of body composition. Measurements of depth of subcutaneous fat, and muscle distribution, by ultrasonic and resistance probes have been used extensively for carcass appraisal (Joblin, 1966), but such measurements are not necessarily satisfactory in predicting total fat and muscle in the animal.

Effects of temperature

The changes in body composition, organ weight, and function seen in other mammals exposed to the cold (Barnett & Mount, 1967) may also be expected to occur in pigs. The increase in growth of hair and the decrease in size of ear observed by Fuller (1965), and the very low skin temperatures of pigs living in the cold (Irving, 1956*a*), may be special adaptations; subcutaneous fat may be increased in the cold (Sørensen, 1962) as an insulative adaptation.

Water

From their observations on rats, Fregly & Waters (1966) suggested that rats in the cold were relatively dehydrated when compared with those in the warm, and that on being placed in a warm environment rats immediately increased their water intakes. This may account for the acute changes in body weight which pigs show when their environmental temperature changes markedly. A change from 20 to 30°C causes a temporary increment in weight, over and above the growth curve, of about 0·5 kg in a 20 kg pig, lasting for several days, and an increment of about 2 kg in a 60 kg pig, lasting for one to two weeks.

Voluntary water consumption is increased in a group of pigs at an environmental temperature of 30°C above that taken at 20°C, by about 40%. This may be in part a thermoregulatory mechanism, although the heat lost by the animal, in warming the water to body temperature, amounts to only about 3% of the total heat loss. The mean daily water intake is approximately 0·1 litre per kg body weight, or 3 litres per kg of meal eaten. Under conditions in which the animals are housed in a pig pen, and fed twice daily at times close to 9 a.m. and 5 p.m., by far the greater part of voluntary water consumption occurs during the period 9 a.m. to 9 p.m. (Holmes & Mount, 1967).

6

CHANNELS OF HEAT EXCHANGE:
I. EVAPORATION

An animal's exchanges of heat with the environment fall naturally into the two main categories of evaporative and non-evaporative (sensible) heat. Whereas evaporative heat loss involves essentially the vaporization of water, sensible heat exchange is characterized by the existence of thermal gradients, and it takes place through the three channels of radiation, convection, and conduction.

When the environmental temperature is reduced, an animal reduces its effective radiating surface area, so that although the radiant heat loss is increased, the total loss is not as great as might be expected from localized measurements on the animal's surface. The same applies to convective heat loss. If the animal is resting on a floor when the temperature is reduced, it changes its posture, rising from a relaxed extended position to one in which it is supported on its limbs in a flexed posture, and this change is associated with a marked reduction in conductive heat loss.

When the environmental temperature rises, sensible heat loss is progressively reduced until it becomes zero when the environmental and body temperatures are equal. Under hot conditions, therefore, an animal's ability to regulate its body temperature depends on an effective evaporative heat loss mechanism.

PARTITIONAL CALORIMETRY

In an extensive series of papers, Winslow, Herrington & Gagge developed the basis of partitional calorimetry in its application to man (Winslow, Herrington & Gagge, 1936*a*, *b*; 1938; Winslow, Gagge & Herrington, 1939; 1940; Winslow & Gagge, 1941; Gagge, 1936; 1940; Gagge, Winslow & Herrington, 1938; Gagge, Herrington & Winslow, 1937). Their method was basically to expose the subject to an environment in which the components responsible for heat exchange by convection and radiation could be varied

131

independently one of the other. They were then able to write heat balance equations for different sets of conditions, and so to arrive at the heat exchange taking place through each of the channels of convection, radiation, and evaporation (conduction played only a small part in their experiments). The system they used to achieve this result is shown in Fig. 6.1; it consisted of an enclosure lined

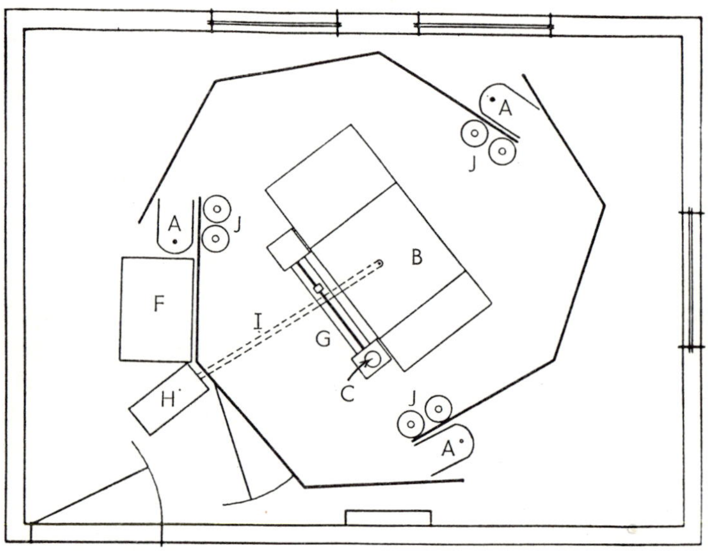

FIG. 6.1. Equipment for the study of thermal interchanges by the method of partitional calorimetry. A, 2500 watt heaters; B, chair; C, aspirating psychrometer; F, chart table; G, platform scales; H, metabolism apparatus; I, hose connection to metabolism apparatus; J, six-inch fans directed to floor of booth (from Winslow, Herrington & Gagge, 1936a, by permission of *American Journal of Physiology*).

by polished copper plates which reflected heat from external radiant heaters on to the subject. The reflected heat then largely determined the subject's radiant environment. The mean radiant temperature was integrated by a hemispherical reflecting convex surface mounted above the enclosure, and this and the subject's mean skin temperature and effective radiating area were taken as the quantities governing radiant heat exchange. Convective heat exchange depended on air temperature and the air movement rate in the vicinity of the subject.

The methods employed in partitional calorimetry as applied to the pig have differed from this, but they have been successful in indicating the relative magnitudes of heat transfer through the different channels. Conductive heat loss is often considerable since the pig spends a variable but often large proportion of time lying down. In this way the pattern of heat transfer is rather different from that of man, although convective and radiant heat losses still account for the bulk of the heat transfer in both cases at temperatures below the zone of thermal neutrality. Another large difference between pig and man arises from the low level of cutaneous evaporative loss which occurs when the pig is exposed to hot conditions in which man sweats profusely.

EVAPORATIVE HEAT LOSS

The loss of heat by evaporation depends on the change of state of water from liquid to gas, a process which takes up heat from the surroundings. It follows that the site of evaporation is important in determining what is going to be cooled in this process. Water evaporating on the surface of an animal's coat takes up heat largely from the surrounding air, and cools the animal only to a very limited degree. In a bare-skinned animal, like pig or man, the evaporation of water on the body surface takes up most of the heat required for the process from the body itself, and so constitutes an efficient cooling mechanism. The actual quantity of heat involved per unit mass of water, whether this is derived exogenously or from sweat, depends not only on the latent heat of vaporization, but also on the cooling of the vapour to ambient temperature, and its expansion to the water vapour pressure of the surrounding atmosphere. The subject is discussed by Hardy (1949), who gives the necessary calculations. The results of these calculations are illustrated for the vaporization of water in the Babraham pen calorimeter at three different temperatures and humidities in Table 6.1. The mean temperatures of skin, expired air, and floor are taken into account in calculating the weighted mean heats of vaporization; the floor is included since it is heat loss from the whole pen which is measured, and not only that directly from the animals.

Evaporative heat loss from the new-born pig has been estimated from the increment in water vapour in the air ventilating a small direct type of calorimeter (Mount, 1962*a*); the theoretical basis of

TABLE 6.1

The total heat of vaporization of water in the Babraham pen calorimeter, including mean values weighted for evaporation from the skin, respiratory tract, and floor for the three temperatures of operation: 9, 20, and 30°C (from Mount et al., 1967, by permission of *Journal of Agricultural Science*)

Calorimeter temperature (°C)	9	20	30
Calorimeter relative humidity	90	60	45
Evaporation at skin surface (kcal/g water)	0·614	0·612	0·612
Evaporation in respiratory tract (kcal/g water)	0·621	0·616	0·611
Evaporation from floor (kcal/g water)	0·595	0·602	0·606
Weighted mean heat of vaporization (kcal/g water)	0·616	0·605	0·609

Values calculated according to Hardy (1949).

such measurements has been discussed by Lasiewski, Acosta, & Bernstein (1966). Precautions were taken to prevent vaporization of water from urine and faeces by collecting these under paraffin oil. Fig. 6.2 illustrates the variation in evaporative and non-evaporative heat losses from new-born pigs over a range of environmental temperature. As a proportion of total heat loss, the

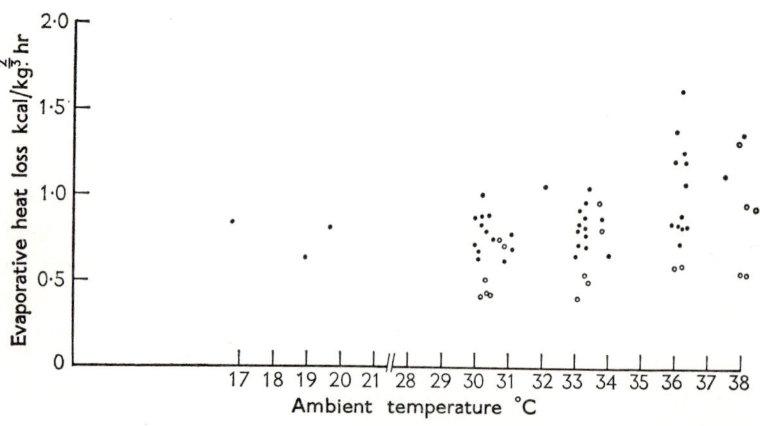

FIG. 6.2. Rates of evaporative and non-evaporative (sensible) heat loss from 22 new-born pigs over a range of environmental temperature. ○, pigs less than one day old; ●, pigs 1 to 7 days old (from Mount, 1962a, by permission of *Journal of Physiology*).

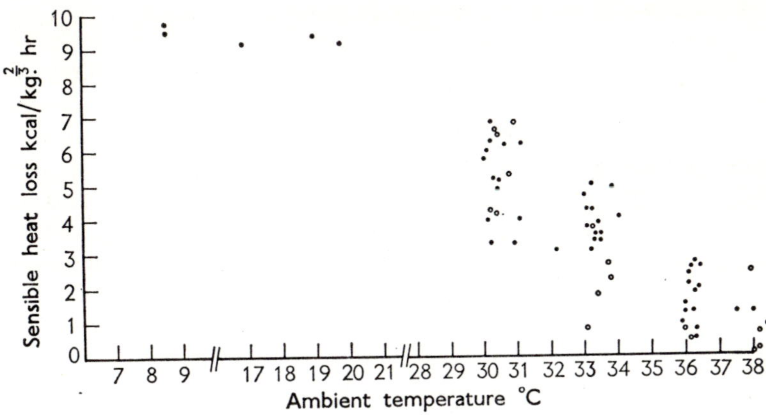

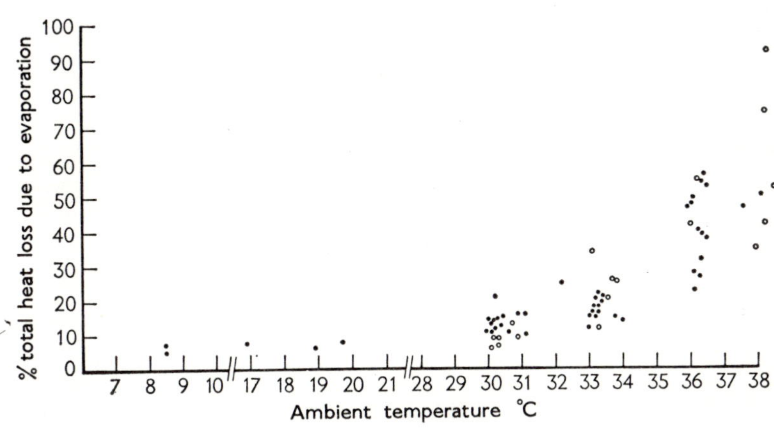

FIG. 6.2.—*continued*

evaporative loss rose from about 8% at 9°C, to 10–20% at 30°C and then to 30–60% at temperatures above the critical level of 34°C. Butchbaker & Shanklin (1964) obtained similar values from measurements made in a gradient layer calorimeter.

Above the critical temperature of 34°C, the new-born pig's respiratory rate increases markedly, but so does rectal temperature (Fig. 6.3). In spite of the rapid onset of polypnoea at high temperatures, therefore, increased ventilation by itself proves to be inadequate as a thermoregulatory mechanism in the pig. Holmes (1966) found a significant ($P < 0.01$) relation between evaporative heat loss and respiratory rate in 20 and 60 kg pigs at 30°C (see Fig. 6.4).

In the lamb, Alexander & Brook (1960) measured respiratory and cutaneous water losses separately over a range of environmental temperature, and found total evaporative losses of the order of four to five times as high as those occurring in the pig (Fig. 6.5). At thermal neutrality in the lamb, the contributions of the respiratory and cutaneous routes are approximately equal, but at higher temperatures the respiratory contribution becomes much larger.

The extra evaporation from the pig at high temperatures in the direct calorimeter experiments could also have come mainly from the increased respiration, making it unnecessary to postulate any

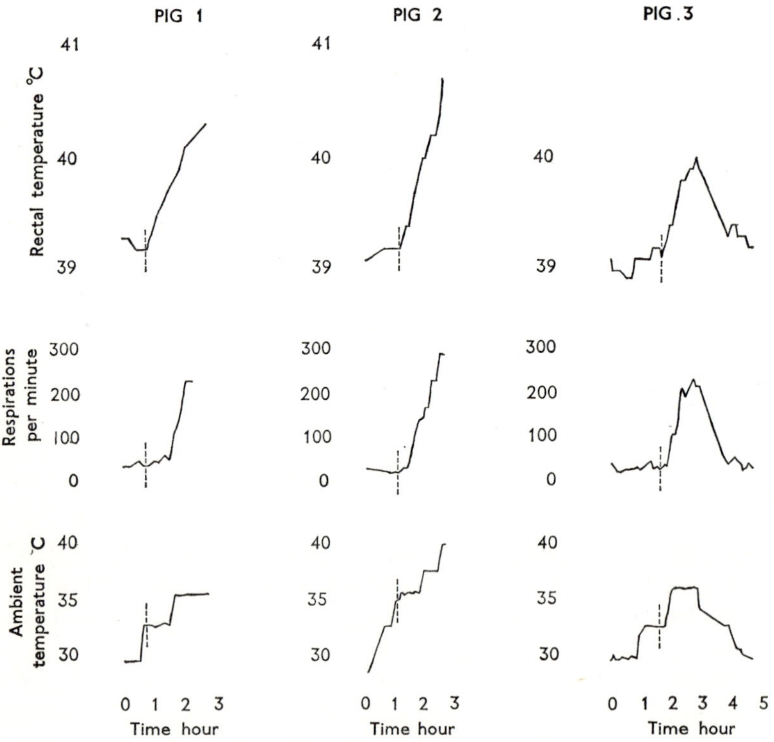

FIG. 6.3. Rectal temperature and respiratory rate in three new-born pigs at ambient temperatures rising above the critical level (pigs 1 and 2), and at temperatures rising above the critical level and then returning to sub-critical values (pig 3).

significant increase in cutaneous water loss. Indeed, if all the increment of water vapour had come from the skin it would still have been indicative of only a low level of ability to sweat. If therefore appears that the new-born pig does not lose heat by evaporation very effectively, with the result that at higher ambient temperatures, approaching the pig's body temperature, heat is stored in the organism and the mean body temperature rises (see Fig. 6.3).

A similarly restricted rate of potential evaporative loss is also characteristic of the older pig. Ingram (1964*b*) found the critical temperature of 3-month-old Landrace pigs to lie between 25 and 30°C as measured by change in the thermal circulation index.

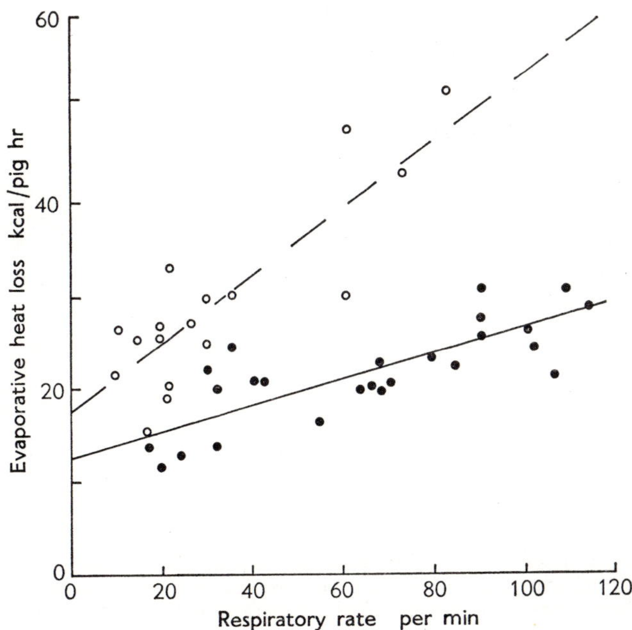

FIG. 6.4. The relation between evaporative heat loss and respiratory rate in pigs at 30°C environmental temperature. ●, 20 kg pigs; ○, 60 kg pigs.

$$20 \text{ kg pigs}: y = 0.138(\pm 0.031) + 12.3; \ r = 0.74^{**}$$
$$60 \text{ kg pigs}: y = 0.357(\pm 0.062) + 16.7; \ r = 0.81^{**}$$
$$^{**}P < 0.01$$

(from Holmes, 1966, by permission of The Queen's University of Belfast).

Using a ventilated capsule placed on the skin, he found that the cutaneous evaporative loss at temperatures above the critical level rose to only about 30 g/m².hr, a figure to be compared with about 60 for the sheep, 70–140 for cattle, and up to extreme limits of 1000 or more for man (Ingram, 1965*a*). In a further study of the

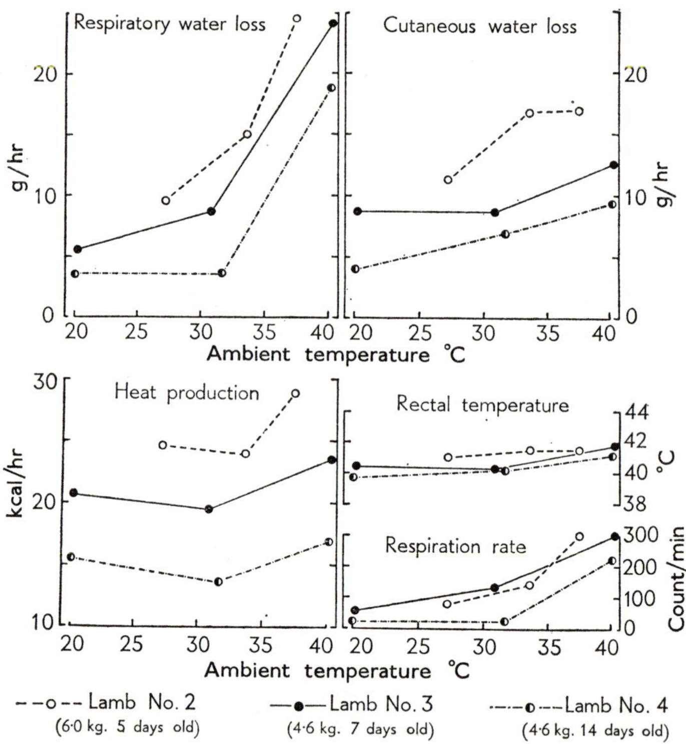

Fig. 6.5. Loss of water by evaporation from the respiratory tract and skin, heat production, and rectal temperature in three unshorn lambs at various environmental temperatures (from Alexander & Brook, 1960, by permission of *Nature*).

effect of humidity on temperature regulation, Ingram (1965*b*) found that an environmental temperature bearing a relation to the deep body temperature of the pig could be expressed as

$$(\text{dry bulb} \times 0\text{·}65) + (\text{wet bulb} \times 0\text{·}35).$$

This weighting gives much less importance to wet bulb temper-

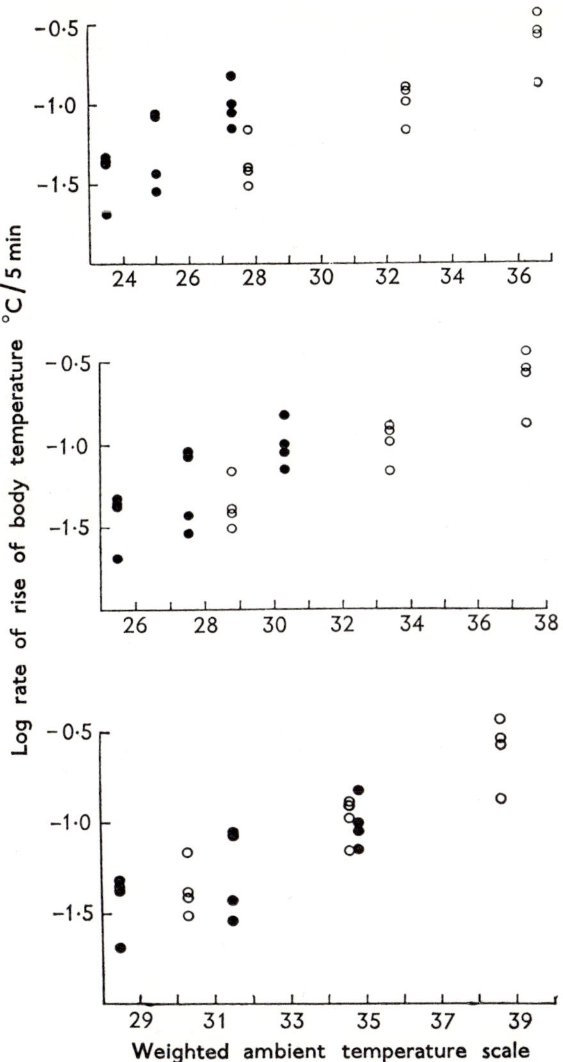

FIG. 6.6. The relation between the logarithm of the rate of rise of body temperature in the pig and three weighted environmental temperature scales, namely:

$$A : DB \times 0\cdot15 + WB \times 0\cdot85$$
$$B : DB \times 0\cdot35 + WB \times 0\cdot65$$
$$C : DB \times 0\cdot65 + WB \times 0\cdot35$$

where DB = dry bulb, and WB = wet bulb. ●, low humidity; ○, high humidity (from Ingram, 1965*b*, by permission of *Research in Veterinary Science*).

ature than has been found necessary for man or cattle. For man (Provins, Hellon, Bell & Hirons, 1962) the appropriate weighting is

$$(\text{dry bulb} \times 0\cdot15) + (\text{wet bulb} \times 0\cdot85)$$

and for cattle (Bianca, 1962) it is

$$(\text{dry bulb} \times 0\cdot35) + (\text{wet bulb} \times 0\cdot65),$$

both of which attach more importance to the wet bulb level than is the case in the pig. The different weightings of wet and dry bulb temperatures as applied to the pig are given in Fig. 6.6.

Although the pig would therefore appear to be unsuited to hot conditions, its wallowing behaviour may greatly increase heat tolerance. To examine this possibility, observations were made on the evaporative loss from the flank of the pig smeared with mud or made thoroughly wet with water (Ingram, 1965a). The rate of water loss from the surface was measured using the ventilated capsule technique. Fig. 6.7 gives the results of such an experiment where a pig was exposed to 35°C dry bulb and 21°C wet bulb, and shows the rate of loss of water vapour from both muddy and wet

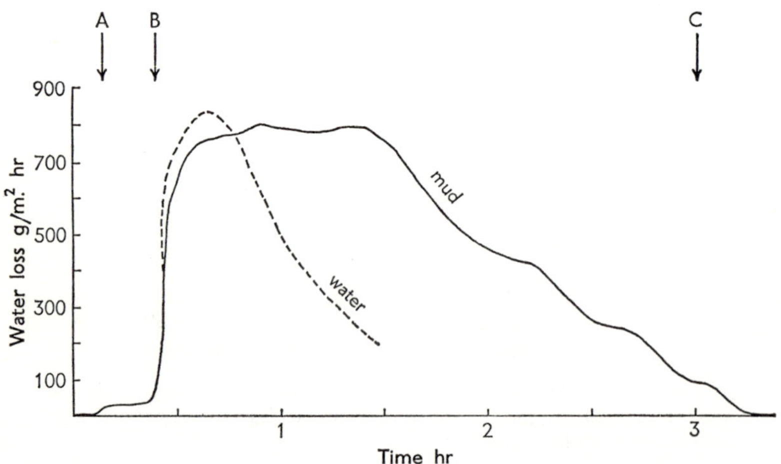

FIG. 6.7. Evaporative water-loss from the skin of a pig measured by the ventilated capsule technique. At A the capsule was placed on the skin. At B mud (continuous line) or water (interrupted line) was smeared over the skin. At C the capsule was removed (from Ingram, 1965a, by permission of *Nature*).

surfaces. In both cases the evaporative rate rose rapidly, but whereas it soon declined with water alone on the pig's skin, evaporation from mud was sustained for several hours, and reached a level comparable with that achieved by sweating in man in a hot environment.

It appears, therefore, that although pigs can make up for the lack of endogenously supplied water for vaporization from the skin by using external water or mud, if this is available, the animals do not actively sweat. Pigs do, however, possess glands, which have been described as apocrine glands, around the base of hair follicles over the general body surface. The ability of these glands to respond to thermal stimulation has been tested using both the ventilated capsule and the starch iodine techniques (Ingram, 1967). No evidence of thermal sweating was obtained, although the glands did respond to adrenaline, noradrenaline, histamine, and 5-hydroxytryptamine. Histological investigation showed that the opening of the duct to the glands was always partly blocked by a plug of keratin. The plug was still present after the glands had been made to discharge in response to adrenaline.

At high temperatures, evaporation necessarily becomes a major avenue of heat loss even in pigs, where evaporative capability is low, since non-evaporative or sensible heat loss is progressively reduced towards zero as environmental temperature rises to the level of body temperature. Morrison, Bond & Heitman (1967) have investigated evaporative losses from two 90 kg Duroc gilts under hot conditions. Cutaneous and respiratory water losses were separated by the animal wearing a mask and being enclosed in a chamber. When the air temperature was changed from 16 to 29°C, at a constant 10°C dew point, the pigs were able to offset the decrease in sensible heat loss by doubling the skin loss and tripling the lung loss (Table 6.2), the latter by means of tripling the pulmonary minute volume (the tidal volume was halved, and the respiratory frequency multiplied by six). The actual rates of vaporization from the skin were similar to those found by Ingram (1964*b*).

A similar investigation of skin and respiratory evaporative losses was made by Brockway, McDonald & Pullar (1965) in sheep. They used a gradient layer calorimeter in which evaporation from the head and from the rest of the body could be measured separately. Whereas cutaneous evaporative heat loss from the body compartment was practically constant at approximately 3 to

TABLE 6.2

Water vapour losses from skin and respiratory tract: mean values from two 90 kg Duroc gilts (from Morrison, Bond & Heitman, 1967, by permission of American Society of Agricultural Engineers)

Environmental temperature °C	Relative humidity %	Water loss g/min per pig		Respiratory rate per min	Tidal volume	Respiratory minute volume, litres	Evaporation from skin, % of total
		skin	resp. tract.				
15	70	0·38	0·28	15	0·79	12	58
29	30	0·72	0·87	86	0·38	33	45
29	50	0·59	1·12	100	0·53	53	35
29	70	0·86	0·78	120	0·50	60	52
29	90	0·83	0·41	143	0·33	47	67

5 kcal/kg per day (corresponding to 8 to 14 g/m².hr for a 60 kg sheep), the respiratory evaporative heat loss (from the head compartment) varied from 2·0 to 17·8 kcal/kg per day (Fig. 6.8). Under warm conditions, therefore, sheep, like pigs, increase evaporative loss through the respiratory tract rather than from the skin. In cattle, however, respiratory evaporative loss rises only

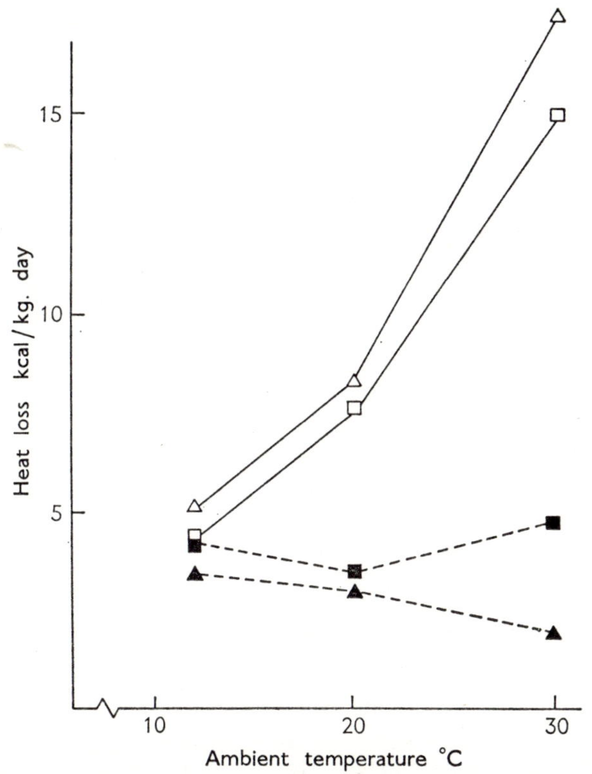

Fig. 6.8. Partition of evaporative loss from two sheep with a constant depth of fleece of 7·5 cm at three environmental temperatures. Continuous lines: respiratory evaporative heat loss; interrupted lines: cutaneous evaporative heat loss (from Brockway, McDonald & Pullar, 1965, by permission of *Journal of Physiology*).

slightly when the ambient temperature is increased between 15 and 40°C, whereas cutaneous evaporation, which is roughly equivalent to the level of respiratory evaporation at 15°C, rises to nearly 85% of the total evaporative loss at 40°C. The effects of environmental temperature on weight losses by respiratory and cutaneous evaporation in Ayrshire bull calves are shown in Fig. 6.9 (McLean, 1963).

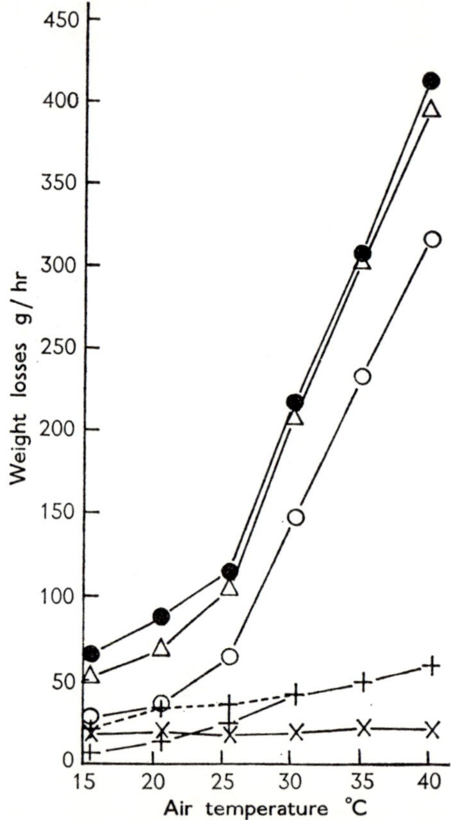

FIG. 6.9. The mean effects of environmental temperature and humidity on the weight losses by respiratory (E_r, +) and cutaneous (E_s, ○) evaporation, on metabolic weight loss (W_m, ×) and on total insensible weight loss (W_t, ●) in Ayrshire bull calves. The interrupted line indicates E_r calculated by the difference method ($W_t - E_s - W_m$). △, $E_s + E_r + W_m$ (from McLean, 1963, by permission of *Journal of Physiology*).

7

CHANNELS OF HEAT EXCHANGE:
II. RADIATION

THE precise computation of the net exchange of heat by radiation between an animal and its surroundings would be an extremely difficult if not impossible task. Fortunately there would be no point in attempting to achieve the precision either of concept or of measurement which is sought by the physicist, since this would be so much out of keeping with the considerable sources of variation which always attend work on living animals. The physical and mathematical basis of the subject of heat exchange by radiation is considered in many text-books; a comprehensive treatment, which permits the selection of the level of approximation appropriate to animal experiments, is given by Jakob (1949, 1957). The subject is presented here, derived from Jakob, so that the relative import-ance of the factors concerned in an animal's radiant heat exchange can be made explicit.

The Stefan-Boltzmann Law for total radiation from a perfectly black body is given by

$$H_R = \sigma A T^4 \tag{1}$$

where H_R = heat transfer rate,

σ = the Stefan–Boltzmann constant, for which the most reliable numerical values are $4 \cdot 96 \times 10^{-8}$ kcal/m^2.hr.°K^4 or $0 \cdot 174 \times 10^{-8}$ B.TH.U./ft^2.hr.°F(abs)^4,

A = effective radiating area of body,

T = absolute temperature of the radiating surface.

This is the one-way radiation from a given body, depending on the fourth power of the absolute temperature T. What is required in practice is the net radiant exchange, which is the difference between the radiant energy leaving the body and the radiant energy entering it from the environment. An often-quoted expression for the net radiant exchange is given by Hardy (1949):

145

$$H = \sigma e_1 e_2 A (T_1{}^4 - T_2{}^4) \qquad (2)$$

where e_1 and e_2 = the emissivities of the surfaces of the body and the surroundings,

T_1 and T_2 = absolute temperatures of body and surroundings.

Emissivity has a maximum value of unity, and this is the value for a perfectly black opaque body, which reflects none of the incident radiation but absorbs all of it. A perfectly opaque reflector, on the other hand, has an emissivity of zero: it reflects all incident radiation and absorbs none of it. Such bodies do not exist in nature; a matt black surface may have an emissivity between 0·95 and 1·0. A nearly perfectly black surface is produced by a hollow enclosure with only a small opening. The opening itself then acts as a surface of absorptivity = emissivity = 1. The proof of this is given by Jakob (1949), and rests on the facts that for an opaque body the sum of absorbed and reflected radiation equals unity, and for a partially transparent body the sum of absorbed, reflected, and transmitted radiation equals unity; and the absorptivity equals the emissivity. Thus an opaque body which absorbs much radiation

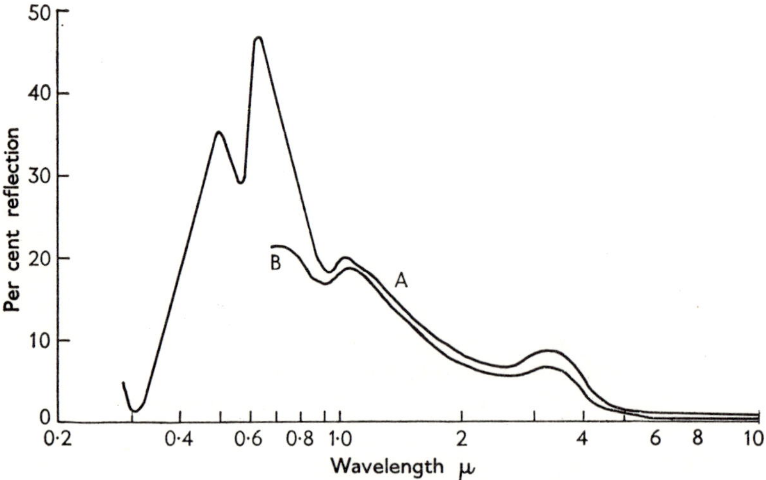

FIG. 7.1. The reflecting power of white and Negro skin over a range of wavelengths of radiation extending from the ultraviolet into the infra-red. A, white skin; B, Negro skin (from Hardy, 1949, by permission of W. B. Saunders Company; modified from Büttner, K.: *Strahlentherapie* **58**, and Hardy, J. D.: *J. Clin. Investigation*, Vol. 13).

reflects little, and vice versa. Emissivity varies with the wavelength of the radiation used; Fig. 7.1 shows the variation in reflection power (and consequently the emissivity) of human skin when the wavelength of radiation is varied. For wavelengths longer than 5μ the emissivity of skin is close to one, and this is true for the majority of surfaces at this wavelength. Table 7.1 gives

TABLE 7.1

Low temperature total emissivities (from the *Handbook of Chemistry and Physics*, 1967, 48th edition, E163, Chemical Rubber Publishing Co.)

Silver, highly polished	0·02
Aluminium, highly polished	0·08
Copper, polished	0·15
Aluminium paint	0·55
Brass, polished	0·60
Black gloss paint	0·90
White lacquer	0·95
Lamp black	0·95

values for the emissivity of commonly occurring surfaces at low temperatures. The emissivity of white lacquer is as high as that of lamp black for these long wavelengths, whereas the shorter wavelengths characteristic of solar radiation are reflected to a large degree by white paint.

Equation (2) is a satisfactory approximation providing that the emissivities e_1 and e_2 both approach unity. As they fall below unity, however, so the error involved in computing the radiant exchange increases, not only on account of emissivity but also because a term involving the areas of the body and its surroundings becomes significant. To avoid these errors, it is necessary to use a more comprehensive equation than eqn. (2), and this requirement is met by Christiansen's equation (Jakob, 1957):

$$H_{\mathrm{R}} = \frac{1}{1 + e_1(1/e_2 - 1)A_1/A_2}\, e_1\sigma A_1(T_1^4 - T_2^4) \qquad (3)$$

where A_1 and A_2 refer to the effective radiating areas of the inner body and the enclosure. This may be re-written:

$$H_{\mathrm{R}} = F\sigma A_1(T_1^4 - T_2^4) \qquad (4)$$

where F is termed the radiative interchange factor, and is defined by eqns. (3) and (4); the factor includes both the configuration and the emissivities of the surfaces. For long concentric cylinders in which the areas of the end-sections are insignificant compared with the areas of the curved surfaces the ratio

$$\frac{A_1}{A_2} = \frac{2\pi r_1}{2\pi r_2} = \frac{r_1}{r_2};$$

for concentric spheres,
$$\frac{A_1}{A_2} = \frac{\pi r_1^2}{\pi r_2^2} = \frac{r_1^2}{r_2^2}$$

(see McGuire, 1953, for a discussion of radiation geometry).

What is important in experiments in which the heat exchange of an animal within an enclosure is considered is that F, the radiative interchange factor, can be determined *for a given situation* without reference to the constitution of the factor, whether it should be represented as $(e_1 e_2)$ or whether it should take the more precise form shown in eqn. (3). The chief disadvantage of this method is that any change in emissivity of the surface of either the animal or of the enclosure would not be recognized, and so would cause error. However, the most probable change in long-wave emissivity would be due to dust or other particulate matter, which would tend to raise the emissivity towards unity. At the long infra-red wavelengths of emission of heat from an animal's skin, the emissivity is already close to unity (see Fig. 7.1), so that any error due to dust would arise only in a polished enclosure. Where the emissivity e_2 of the enclosure surface approaches unity, the value for F given by eqns. (3) and (4) tends towards e_1.

The wavelength of maximum emission of radiation from a surface, λ, is given by Wien's displacement law (Jakob, 1949), which states that $\lambda_{max}.T = 0{\cdot}288$ cm°K, where T is the absolute temperature. For a skin temperature of 30°C, λ_{max} is therefore $0{\cdot}288/303$ cm $= 9{\cdot}5\mu$; for 35°C it is $9{\cdot}35\mu$.

Heat exchange by radiation in animals is conveniently considered in two stages. The first of these deals with exchange when radiation from the surroundings is all long-wave, that is, derived from surfaces at a range of temperatures extending from several hundred degrees centigrade downwards (a wavelength of radiation of 5μ corresponds to the surface at about 300°C). The second stage goes on to include the effects of shorter wavelengths, including the visible spectrum and ultra-violet. Fig. 7.2 shows the natural division which occurs between long-wave and short-wave radiant flux in the wavelength region of 3μ.

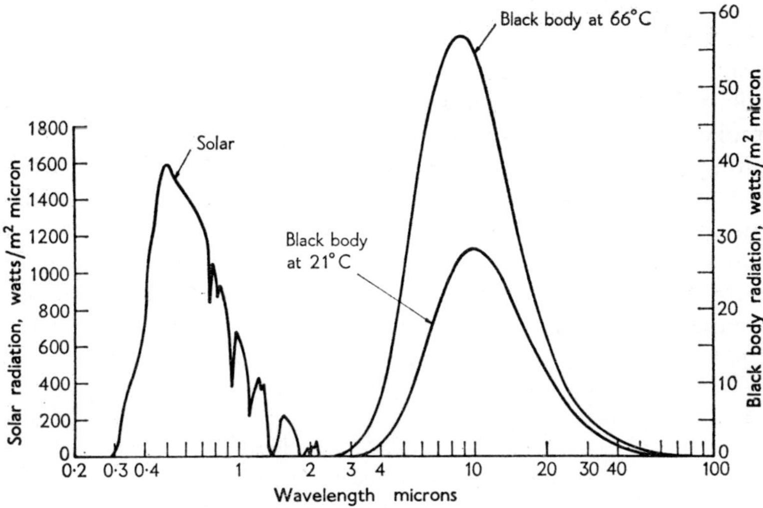

Fig. 7.2. Solar radiation at the ground for a zenith sun, and radiation from black bodies at 21 and 66 °C (from Bond, Kelly, Morrison & Pereira, 1967, by permission of American Society of Agricultural Engineers).

LONG–WAVE RADIANT EXCHANGE

An important feature of long-wave radiation is that it does not pass through the majority of substances which are transparent to the shorter wavelengths of the visible spectrum. A common practical example of this is the horticultural glasshouse. Glass allows the passage of the shorter wavelengths of radiation from the sun and sky, so that energy enters the house and warms it. The warm surfaces inside the house emit heat at a much longer wavelength of radiation to which glass is opaque, with the result that direct radiation from objects inside the house to the outside is prevented; the net inward movement of radiant energy produces a raised equilibrium temperature inside the house, with heat loss dependent on ventilation, and on radiant and convective loss from the surface.

Long-wave radiant exchange and surface temperature

A somewhat similar situation can arise in a hospital incubator for babies, since Perspex (polymethyl methacrylate) is similarly

opaque to long-wave radiation. The consequences for the baby's
thermal environment have been investigated by Hey and Mount
(1967). Radiant heat exchange takes place between the baby on the
one hand and, on the other hand, the inner surface of the Perspex
wall and the mattress on which the baby lies. The temperature of
the wall therefore becomes important in determining the baby's
radiant heat loss, since it occupies the major part of the solid angle
subtended at the baby. The wall temperature lies approximately
45% down the gradient between the temperatures of the incubator
air and the room air, so that in a cool room the baby's radiant heat
loss can be considerable. A thin Perspex shield placed between the
baby and the wall reduces this loss. The shield tends towards the
temperature of the incubator air, and since it is opaque to long-

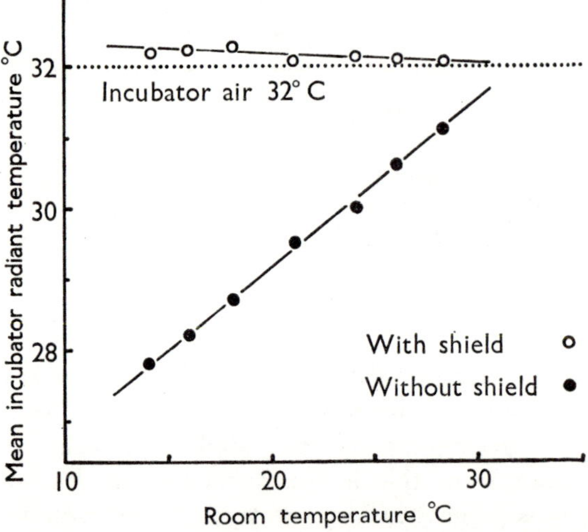

Fig. 7.3. The effect of room temperature on the mean radiant
temperature within an incubator, with and without an internal
shield at the temperature of the incubator air (from Hey & Mount,
1966, by permission of *Lancet*).

wave-radiation from the baby, it acts as a warmer radiant environment than the incubator wall (Fig. 7.3). The use of such a radiant
shield significantly lowers the baby's oxygen consumption rate
from about 9·0 to 7·6 ml./kg.min (mean result from measurements
on six babies), and its effect on skin temperature is shown in
Fig. 7.4.

Work on the incubator involved the difficulty of obtaining accurate surface temperature measurements. In Fig. 7.5, the temperatures of the inner and outer surfaces of the wall, and the temperature of the mattress, are given both as surface thermocouple and as radiometer measurements. The two methods give similar results for the mattress, which is close to air temperature, but for the wall, which is warmer than the room air on the external

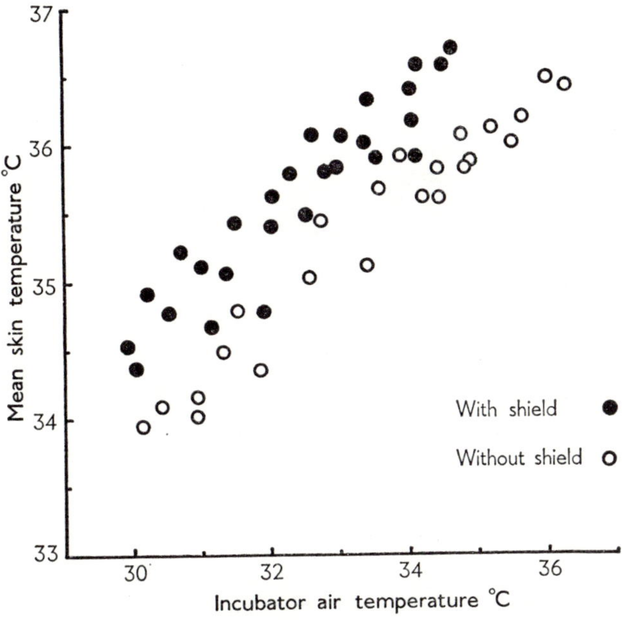

FIG. 7.4. The effect of a radiant shield on the mean skin temperature of 13 healthy human infants in an incubator, over a range of incubator air temperatures. Room temperature: 20–22°C (from Hey & Mount, 1967, by permission of *Archives of Diseases in Childhood*).

side, but cooler than the incubator air on the internal side, there is the well-recognized deviation of the surface thermocouple reading towards air temperature (Molnar & Rosenbaum, 1963). The deviation from the radiometer reading is most marked in the case of hard surfaces, as distinct from results on indentable surfaces such as skin. The skin temperatures of pigs, referred to in this chapter and elsewhere, have been measured by thermojunctions

F

under thin rubber patches; the difficulties in obtaining reliable measurements of skin temperature are discussed by Findlay (1950) and Vere (1958).

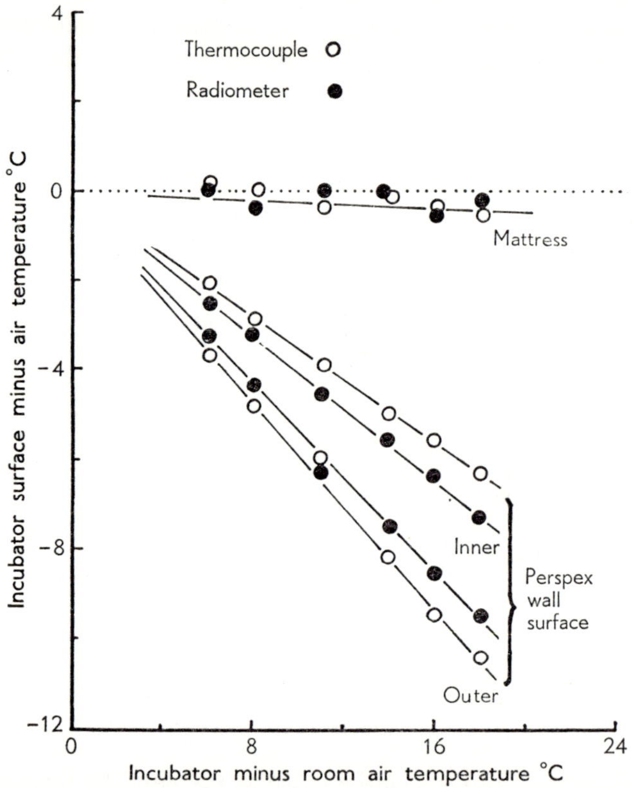

FIG. 7.5. Thermocouple and radiometer estimates of the relation between incubator air and mean surface temperatures in an incubator (from Hey & Mount, 1967, by permission of *Archives of Diseases in Childhood*).

The use of polyethylene

Polyethylene is one of the few substances which transmit the long infra-red waves, and in Fig. 7.6 the percentage transmittance of glass, Perspex and polyethylene sheet are given in relation to wavelength. Glass and Perspex transmittance fall towards zero as the wavelength increases beyond 2–5μ, but the transmittance of polyethylene remains high, with sharp troughs recurring at regular wavelength intervals due to the interference pattern associated with the CH_2 grouping in the molecule. If polyethylene is used instead of glass in making a greenhouse, the advantage that glass confers of preventing direct radiant heat loss from inside is not retained. The value of the polyethylene, in common with glass, would then lie in providing the insulation associated with boundary layers of air and so preventing some of the heat loss by convection.

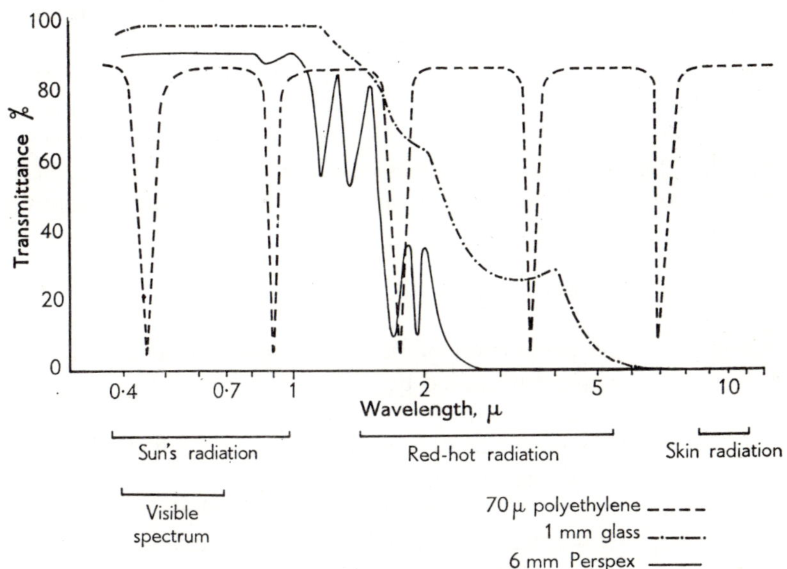

FIG. 7.6. The percentage transmittance of polyethylene, glass, and Perspex for radiation extending from the visible spectrum into the infra-red (partly diagrammatic, for clarity).

For radiant heat emitted by the skin, transmittance is at a minimum through glass and Perspex, but high through polyethylene. The property of polyethylene of transmitting such long-wave radiation has been used in shielding a radiometer (Funk, 1959), in experiments on the radiation cooling of dairy cattle carried out by Shanklin & Stewart (1958), and in the determination of the radiant heat exchange of the new-born pig (Mount, 1964*b*).

In the last application, the animal was housed in a cage inside a double-walled tent of thin polyethylene sheeting, 37μ thick, which itself was inside a metal chamber in a water-bath. The chamber wall temperature was determined by the water temperature; the temperature of the air around the pig inside the tent could be raised by an electric heater, from which the pig was shielded, to a level above that of the wall and its immediate vicinity. Thus the polyethylene allowed the separation of the temperatures of the convective and radiant environments, since it was largely transparent to radiant exchange between the pig and the metal chamber wall, but prevented any controlling effect of the temperature of the water-bath on that of the air around the pig.

In this situation, equations based on the Stefan–Boltzmann Law (using eqn. (2), since the emissivities approach unity) can be written for the radiant heat exchanges between the pig and the components of its environment, including the cage, the inner and outer polyethylene layers, and the chamber wall. An expression can then be derived for an effective mean radiant environmental temperature which takes these various components into account; this is T_2 in eqn. (2):

$$T_2{}^4 = (1-p)\,T_A{}^4 + p[(1-m)\left(T_{p_1}{}^4 + mT_{p_2}{}^4\right) + m^2 T_W{}^4] \qquad (5)$$

where p = the proportion of the solid angle, subtended at the pig, not occupied by cage structure,

     T_A = air temperature,

     m = transmittance of one 37μ layer of polyethylene at wavelength 9–10μ,

     T_{p_1} = temperature of inner layer of polyethylene,

     T_{p_2} = temperature of outer layer of polyethylene,

     T_W = temperature of chamber wall.

The cage structure and the shield between the pig and the air heater inside the polyethylene tent were assumed to be at air temperature, hence the term $(1-p)T_A{}^4$ which contributes to the effective mean radiating temperature.

T_1 in eqn. (2) is the pig's mean skin temperature. This presents a problem in estimation; the skin temperature varies over the animal's surface, and the weightings to be attached to the different temperature areas clearly depend on the animal's situation and posture which affect the degrees to which the different areas take part in radiant heat exchange. As an approximation, ear temperature was used as representative of ear and limb temperature, and back temperature as representing trunk and head. Measurements made on the 2 kg pig by K. F. Hosie (unpublished observations) showed that the ears and limbs together constitute 27% of the total body surface area, and head and trunk 73%. The mean skin temperature could therefore be represented as (27 ear + 73 back)/ 100. For radiant exchange purposes, however, this takes no account of the obscuring of the solid angle subtended at some parts of the pig by other parts of the animal. From postural considerations it seemed likely that under cool conditions a better weighting would be (16 ear + 84 back)/100, while with the more relaxed posture found under warm conditions the weighting would become (20 ear + 80 back)/100. In fact, use of either of the last two weightings for calculations of results at either 20 or 30°C made only $\pm 1\%$ difference to the determination of the animal's effective radiating area, and no measurable difference to the radiant-convective partition. Ear and back skin temperatures were

TABLE 7.2

Mean rectal temperature, and mean skin temperature from different weightings of ear and mid-back temperatures, at given air and effective wall temperatures, for six pigs (from Mount, 1964*b*, by permission of the *Journal of Physiology*)

Air temp. (°C)	Effective wall temp. (°C)	Rectal temp. (°C)	Skin temp. (°C)		Mean skin temp. (°C)		
					16 ear + 84 back	20 ear + 80 back	27 ear + 73 back
			Ear	Mid-back	back	back	back
30·0	29·2	39·4	35·8	37·3	37·1	37·0	36·9
30·0	19·4	39·4	32·9	35·9	35·4	35·3	35·1
20·2	19·3	39·6	29·4	34·0	33·3	33·1	32·8
20·2	10·0	39·6	25·4	32·6	31·5	31·2	30·7

measured on pigs in the chamber at different air and effective wall temperatures, and the effects of various weightings on estimates of mean skin temperature are given in Table 7.2.

In these experiments, the unknown quantities in eqn. (2) were the net radiant heat loss, H_R, e_1, e_2 and A, the pig's mean effective radiating area. The product $(e_1 e_2)$ may be regarded as a radiative interchange factor in this given situation (see eqn. (4)). This factor was determined experimentally from measurements on heated spheres in the cage at different combinations of air and effective wall temperatures. Two solid matt black spheres, each containing an electric heater winding, and of radii 1·27 and 1·97 cm, were suspended in the chamber and maintained at 35°C. At equilibrium, the heat loss equalled the heat input. The spheres were of duralumin, so that their surface temperatures were the same as those recorded by thermocouples in their centres. The surface area of each sphere could be calculated, but the suspending wires presented a complication since they would be warmed by conduction and would therefore lose heat to the surroundings to an unknown degree. The use of two spheres was designed to overcome this difficulty. Each sphere had similar suspensions, so that the difference in heat inputs required to maintain the sphere temperatures could be related to the difference between the measured areas of the spheres, the effective areas of the two suspensions cancelling out. With the air temperature and air movement held constant, the effective wall temperature was lowered; concomitantly the electrical heat input to the spheres was increased to maintain the sphere temperatures constant. Thus the difference between the increments in heat inputs could be related to the fall in mean radiant temperature and to the difference between the areas of the spheres to give the value of the radiative interchange factor, $F = e_1 e_2$:

$$F = \frac{\Delta H_\mathrm{L} - \Delta H_\mathrm{S}}{\sigma(A_\mathrm{L} - A_\mathrm{S})(T_1{}^4 - T_2{}^4)}$$

where $\Delta H =$ increment in heat input,

$\qquad A =$ surface area,

$\qquad T_1 =$ initial mean radiant temperature,

$\qquad T_2 =$ final mean radiant temperature,

$\qquad \sigma =$ Stefan–Boltzmann constant,

and the suffixes L and S refer to the large and small spheres respectively.

The reference base is thus taken as the difference between the spheres, so avoiding the mensuration difficulty associated with the use of only one sphere.

The value of $(e_1 e_2)$ was taken to be the same for the pigs as for the spheres, where it approximated to 0·98. This is close to the value of Hardy's (1939) determination of the emissivity of human skin between 0·97 and 0·99. The radiative interchange factor of $(e_1 e_2)$ takes account of the physical characteristics and geometrical arrangement of the chamber in measurements on both the spheres and the pigs; differences in configuration between the spheres and the animal do not matter since both e_1 and e_2 approach unity. With the form of experiment used, (e_1e_2) did not enter the calculation of radiant heat loss, but it did affect directly the estimate of the effective radiating area. From the measurement of metabolic rate of a pig in the cage, and from knowledge of the probable evaporative heat loss (see Chapter 6), the radiant heat loss, the effective radiating area, and the heat lost by convection could be calculated (for the details of formulation, calculation, and results, see Mount, 1964*b*).

The results of these experiments showed that at 20 and 30°C environmental temperatures the radiant and convective heat losses of new-born pigs were approximately equal (see Table 7.3).

TABLE 7.3

The proportions of non-evaporative heat lost by radiation and convection from the 2 kg pig (from Mount, 1964*b*, by permission of the *Journal of Physiology*)

Air temp. (°C)	Effective wall temp. (°C)	Radiant heat loss (kcal/hr)	Convective heat loss (kcal/hr)	Total non-evaporative heat loss (kcal/hr)	Radiation (%)	Convection (%)
29·8	29·3	4·88	3·36	8·24	59	41
30·1	19·2	9·12	2·76	11·88	77	23
20·3	19·4	7·32	7·32	14·64	50	50
20·1	9·7	10·18	6·96	17·14	59	41

At 20°C the convective loss increased above that expected: experiments with a vibrating sphere suggested that this might have been due to the shivering which occurs in new-born pigs at 20°C. The effective radiating area of the pig increased from 67% to 76% of its total surface area as the temperature rose from 20 to 30°C, suggesting a postural change. The total surface area was estimated from Brody's (1945) formula, $0.097W^{0.633}$, which gives the surface area in m^2 for a pig of weight W kg. It was found that when the environmental temperature fell from 30 to 20°C, the combined reductions in mean skin temperature and effective radiating area brought about a rate of heat loss of about two-thirds of what it would have been without these reductions. The fall in mean skin temperature had approximately twice the effect of the accompanying decrease in effective area in reducing heat loss. Wall and air temperature changes were found to be approximately equivalent to each other in their effects on heat loss under conditions approaching those of still air. A fall in wall temperature of 1°C, at constant air temperature, increased the metabolic rate at equilibrium by 0·30 kcal/hr; at constant wall temperature, a fall of 1°C in air temperature increased the rate by 0·28 kcal/hr. Koch (1962) found from subjective comfort tests on lightly dressed persons that an increase of 1°F in air temperature was compensated by a decrease of 1·39°F in the mean radiant temperature, when the air velocity was less than 10 cm/sec (20 ft/min), conditions comparable with those in the pig experiments. The similarities between the results on man and pig suggests that the subjective comfort estimate is related to the thermal demand imposed on the organism by the environment.

Another approach to the separation of radiant from convective heat transfer, in sheep, was adopted by Joyce, Blaxter & Park (1966). They exposed two sheep in two respiration chambers in which the air velocities and their distributions were similar, but which had different wall emissivities in the infra-red. Varying the emissivity changed the effective black-body temperature of the walls to which the animals were losing heat, and in each set of circumstances an appropriate radiative interchange factor could be determined. From the results of these and other experiments, a prediction equation was provided which gave good agreement between measurements made in the laboratory and other measurements made under a variety of conditions out of doors.

FULL-SPECTRUM RADIANT EXCHANGE

The long-wave radiant energy exchange of an animal out of doors is with the ground and the atmosphere, and the animal receives short-wave radiant energy from the sun both directly and as a diffuse solar radiation due to scattering in the atmosphere.

Kelly, Bond & Heitman (1954) give the direct solar energy, in the wavelength region of 0·4 to 3μ, as 610 kcal/m².hr, and the atmospheric radiation (including diffuse solar radiation), in the region of 3 to 50μ, as about 320 kcal/m².hr. They made measurements with a directional radiometer of the radiant energy expected to fall on an animal from the earth and sky hemisphere both under a shade and when exposed to the sun. The reduction in radiant intensity due to standard cattle shades in California was found to be 50–65% on the animal's back from the sky hemisphere and 25 – 30% from the earth hemisphere.

Lee & Vaughan (1964) used the physiological reactions (rectal temperature, pulse rate, and sweating) of man under desert conditions to establish the increment of air temperature equivalent to the heat load due to solar radiation. At an air temperature of about 45°C, they found that an increment in air temperature of approximately 8°C was required to produce the same effect as solar radiation. In other experiments, Gagge, Stolwijk & Hardy (1965) exposed unclothed human subjects to a variable source of thermal radiation, and measured evaporative loss continuously while the ambient temperature was varied over the range 15–30°C. The subject was allowed to choose a heater setting necessary for comfort, and in this way the man was used as a radiometer. As a result, it was possible to determine heat exchange by radiation in a complex radiant environment, and to derive an environmental temperature scale which is useful for comparing physiological responses under widely different physical conditions. It should be possible to determine a similar type of scale for an animal by putting a pig, for example, in a multiple choice situation where it is free to move into zones characterized by different levels of radiant and convective heating.

Skin colour

Fig. 7.1 shows that the colour of the skin of man has no apparent effect on emissivity and reflectivity in the long wavelength zone

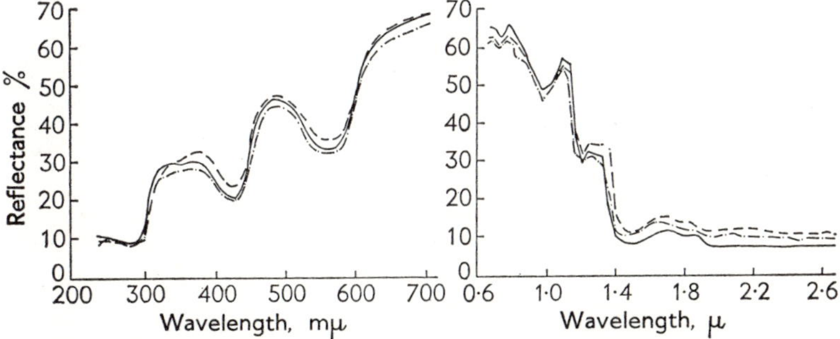

FIG. 7.7. The mean spectral reflectance values of the skin of three Chester White pigs (from Kuppenheim, Dimitroff, Melotti, Graham & Swanson, 1956, by permission of *Journal of Applied Physiology*).

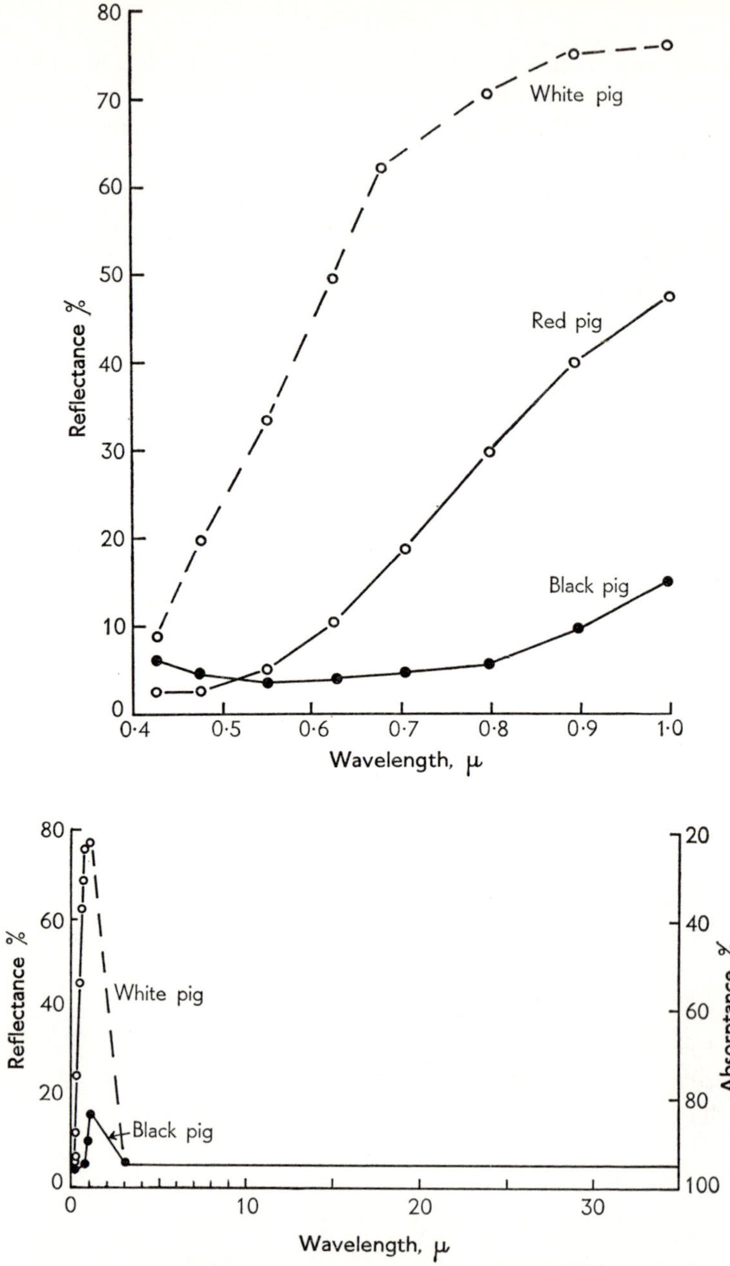

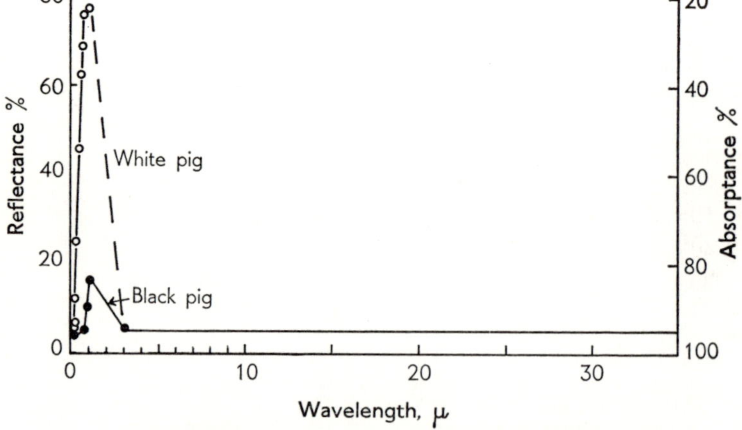

FIG. 7.8. The spectral reflectance of pig skin. Upper diagram: white, red, and black pig skin, between wavelengths of 0·4 and 1·0μ; lower diagram: reflectance and absorptance of white and black pig skin between wavelengths of 0·4 and 35μ (from Kelly, Bond & Heitman, 1954, by permission of *Ecology*).

of radiation; in this connection there is a general similarity between pig and human skin. In the reflectance of solar or short-wave radiation, however, the colour of the skin or coat is important. The spectral reflectance of Chester White pigs is given in Fig. 7.7 for wavelengths from o·2 to 2·6μ, that is through the visible spectrum and into the infra-red. From a wavelength of about 1·5μ and longer the reflectance falls to rather less than 10%.

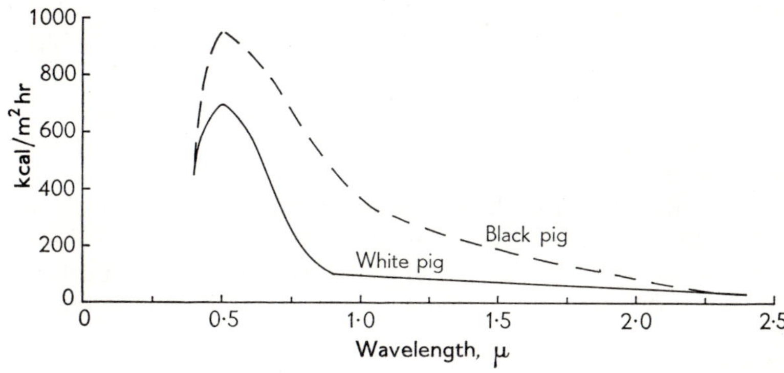

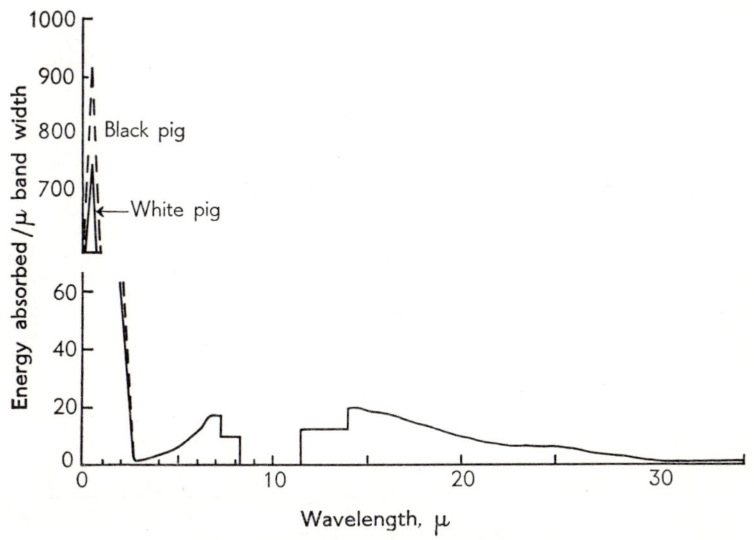

FIG. 7.9. The solar and atmospheric radiant energy absorbed by black and white pigs for typical atmospheric conditions at Davis, California. Upper diagram: absorption over the wavelength range o·4 to 2·4μ; lower diagram: absorption up to 35μ (from Kelly, Bond & Heitman, 1954, by permission of *Ecology*).

Values for the spectral reflectance of white, black, and red pig skin for wavelengths of 0·4 to 1·0μ, determined spectrophotometrically by other workers, are given in Fig. 7.8, together with a diagrammatic extension to a wavelength of 35μ. Fig. 7.9 gives the solar energy absorption by white and black pigs; the absorption curves were obtained by multiplying radiation intensity by absorptance at given wavelengths. The white pigs reflect 51% of solar energy between 0·4 and 1μ while black pigs reflect only 7%. The total energy absorption over the range of 0·4 to 3μ is 331 kcal/m².hr for the white pig and 629 for the black pig, so that the black pig absorbs 90% more energy than the white pig between these limits (Kelly *et al.*, 1954). Over the whole range from 0·4 to 50μ, the white pig absorbs 585 kcal/m².hr and the black pig 883, only 50% more than the white pig.

Although white pigs have this advantage, the two animals are similar in the long-wave zone, in which they both have emissivities close to unity, so that the overall heat tolerances of the animals are not so very different. Indoors, there would be no difference

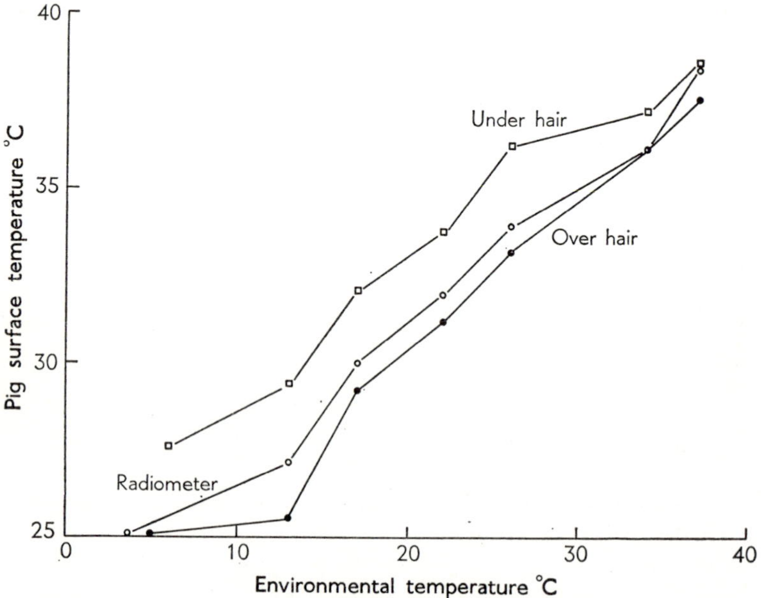

FIG. 7.10. The surface temperatures of pigs, at environmental temperatures from 4 to 37°C, determined by touch thermocouple and by radiometer (from Kelly, Bond & Heitman, 1954, by permission of *Ecology*).

between white and black pigs, save for a small effect with bright heaters. The relatively sparse hair coat of the pig has some effect on the animal's mean radiant temperature. This is shown in Fig. 7.10 where thermocouple and radiometer temperatures from pigs are compared over a range of environmental temperature; the radiometer 'sees' a combination of hair and skin temperatures, and consequently records a value which lies between the two.

In cattle, Stewart (1953) found a range of values of spectral reflectance of cattle coats for radiation between 0·3 and 1·2μ wavelength (Fig. 7.11), with results similar to those on pigs. A number of authors have considered the effect of cattle hair-colour on environmental adaptability (Bonsma, 1943; Rhoad, 1940).

Measurements of the emissivity of the skin and the incident radiant heat load per unit area give an indication of the total heat load absorbed at the animal's surface. A lot of this may be dissipated by convection at the surface, so that no additional burden is placed on the organism's other channels for heat loss. If, however, the radiant energy penetrates into the tissues for any distance,

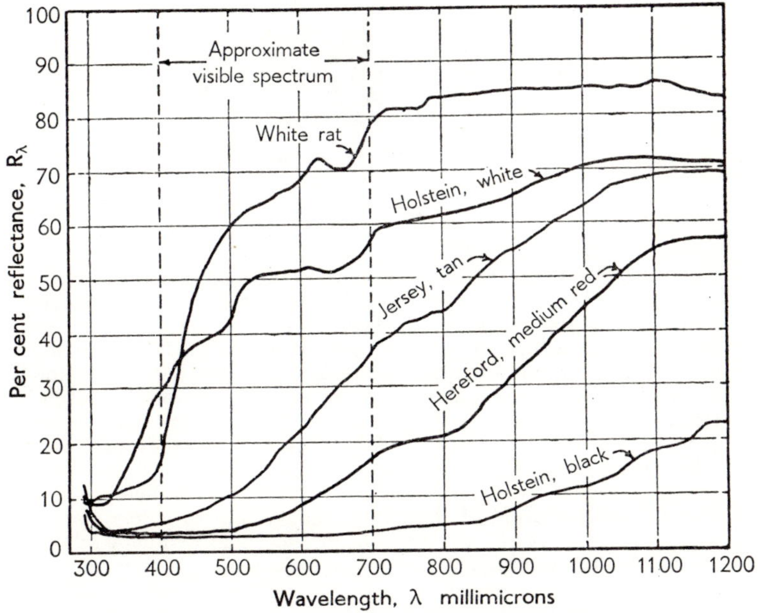

FIG. 7.11. The percentage reflectance of hair coats of various cattle and the white rat, over a range of wavelength of radiation (from Stewart, 1953, by permission of *Agricultural Engineering*).

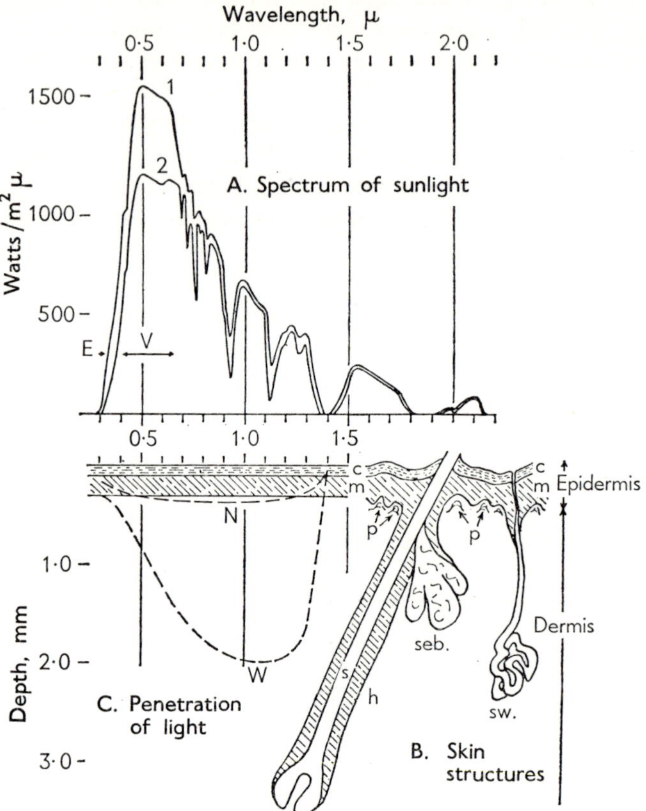

FIG. 7.12. A—Spectral distribution of sunlight at earth's surface. 1, sun at zenith; 2, sun at 60° (four hours) from zenith. V, spectral limits of human vision; E, spectral limits (within sunlight) for sunburn, antirachitic action, and cancer induction. B—Diagrammatic representation of skin structures. A schematized conception in which the dimensions should not be taken as generally representative, since the skin may vary widely in its thickness; c, corneum (horny layer of epidermis); m, Malpighian layer of epidermis; sw., sweat gland; seb., sebaceous gland; p, the most superficial blood vessels, arterioles, capillaries, and venules; h, hair follicle; s, hair shaft. C—Penetration of light into human skin as a function of wavelength. Curves N and W indicate Negro and white skin and are rough estimates of depths at which radiation of the corresponding wavelengths is reduced to 5 per cent of its incident value. The curves are based on the results of: Hardy & Muschenheim (1934) *J. clin. Invest.* **15**, 1; Kirby-Smith, Blum & Grady (1942) *J. nat. Cancer Instit.* **2**, 403; Thomson (1955) *J. Physiol.* **127**, 236 (from Blum, 1961, by permission of *Quarterly Review of Biology*).

TABLE 7.4

The temperatures of various surfaces exposed to bright sunshine, expressed as excess above air temperature to nearest 0·5°C (from Landsberg, 1964, by permission of Elizabeth Licht)

Natural surfaces	
Soil surface under 5 cm grass	0·1
Intermittently shaded soil surface	1·0
Tree leaves	2·0
Short grass tips	6·0
Bare sand	9·5
Bare rock	11·0
Bare dark soil	15·5
Road surfaces	
Gravel drive	8·0
Concrete	17·5
Asphalt	22·5
Roofing materials	
Asphalt shingles	17·0
Dark slate	18·0
Asbestos shingles	19·0
Galvanized metal	
fresh aluminium paint	6·5
unpainted	16·0
red paint	19·5
black paint	26·0
Structural materials and parts	
Brick steps	6·5
Brick wall, unpainted	10·5
Wooden siding, red paint	8·5
Window surface	7·0
Wood block	
white lead paint	11·0
pink paint	11·0
yellow ochre paint	14·5
red oil paint	15·5
soot covered	17·0
Cork	
white oil paint	7·5
black oil paint	28·0
Stone	
white oil paint	6·5
black oil paint	18·0

heat is taken into the blood and must necessarily be dissipated else-where than at the site of absorption, either by sweating, or as non-evaporative heat loss. In this connection, measurements on man may demonstrate one reason why the man with light skin is more subject than the man with dark skin to heat stroke from direct solar radiation. Fig. 7.12 shows the penetration of light and dark skin by the sun's radiation, and the far greater penetration achieved through light skin. The much greater absorption of energy which occurs at the surface of the skin in the man with a dark skin could lead to a greater local dissipation of heat. Thus although dark skin absorbs more total energy from direct solar radiation than does light skin (see Fig. 7.1), local heat loss at the surface may more than compensate for this. In this connection, Thomson (1955) con-cluded from his studies of the stratum corneum of Europeans and Africans that the protection offered by pigment in the skin against ultraviolet radiation more than compensates for the disadvantage of dark skin in absorbing more radiant energy from the sun. Table 7.4 shows how surfaces of various materials differ from each other in temperature when they are exposed to bright sunlight.

8

CHANNELS OF HEAT EXCHANGE: III. CONVECTION AND CONDUCTION

HEAT EXCHANGE BY CONVECTION

WHEREAS radiant heat exchange depends on the radiant temperature and effective radiating area and emissivity, heat exchange by convection depends on the surface temperature of the body and its shape, surface characteristics, and size, and on the air temperature and the air movement rate which impinges on the body.

NATURAL AND FORCED CONVECTION

The basis of convective heat exchange has been discussed by Hardy (1949). It is a form of heat transfer which depends on the redistribution of molecules within a fluid, as distinct from the conduction of heat in which there is no actual translocation of molecules. Natural convection occurs as the result of a temperature difference in a fluid; the way in which air rises after being warmed by a hot body is an example. Forced convection implies that the fluid movement takes place not as a result of temperature gradients but as the result of an external force moving the fluid; an example is cooling by air blown by a fan. In nearly still air, a boundary layer forms on the surface of an object, and it is this layer which produces, for example, the major part of the thermal insulation across a thin wall, the substance of which has a high thermal conductivity. Fig. 8.1 shows the temperature gradients across the Perspex wall of an incubator, and demonstrates that the temperature change is nearly all in the boundary layers. Movements of the surface, as in the 'shivering sphere' referred to in Chapter 7 in connection with radiant exchange, and in an active animal, will increase convective heat exchange considerably by disrupting the boundary layer; forced convection has the same effect. The nature of the surface of the body, that is whether it is smooth or indented, naked or coated, has a considerable effect on convective exchange; the size also has an effect, the smaller the radius of a part the greater being the convective heat loss per unit area for a given body—ambient temperature difference. These effects can be combined to give a 'characteristic dimension' for the object in question, and Hardy (1949) points out that the adult human body behaves like a cylinder 7 cm in diameter, or a sphere 15 cm in diameter, if everything but convection is neglected.

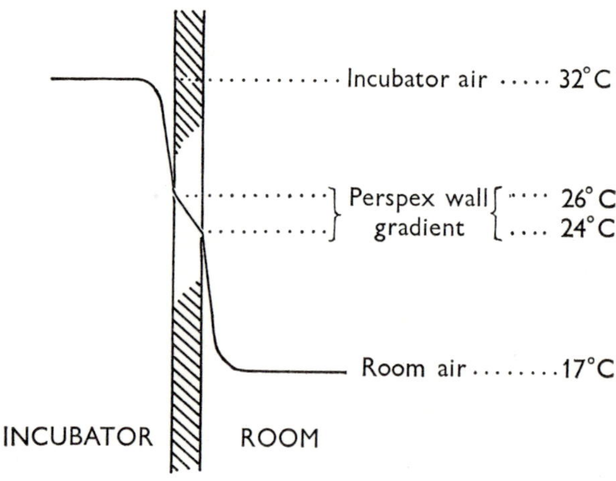

Fig. 8.1. Temperature gradients across the Perspex wall of a heated incubator; the thickness of the boundary layers is not drawn to scale (from Hey & Mount, 1966, by permission of *Lancet*).

Observations on new-born pigs (Mount, 1964*b*) have shown that at low air movement rates heat losses by radiation and convection are approximately equal. The same was found for larger pigs by Bond, Kelly & Heitman (1952). The comparative effects of radiant and convective heat exchanges on three different types of body are illustrated in Fig. 8.2. The globe thermometer, with no thermoregulatory mechanism, shows a marked rise in temperature in full sunlight. The leaf's temperature, however, closely follows that of the surrounding air as a result of the proportionately large convective loss due to its high surface-area/volume ratio and its movement, and as a result of transpiration. The cow's deep body temperature remains at a steady level in consequence of thermoregulatory mechanisms involving increased evaporative heat loss, and increased non-evaporative losses due to a higher skin temperature.

The estimate of air velocity at the body surface is difficult to obtain for the animal as a whole, particularly in a relatively linear

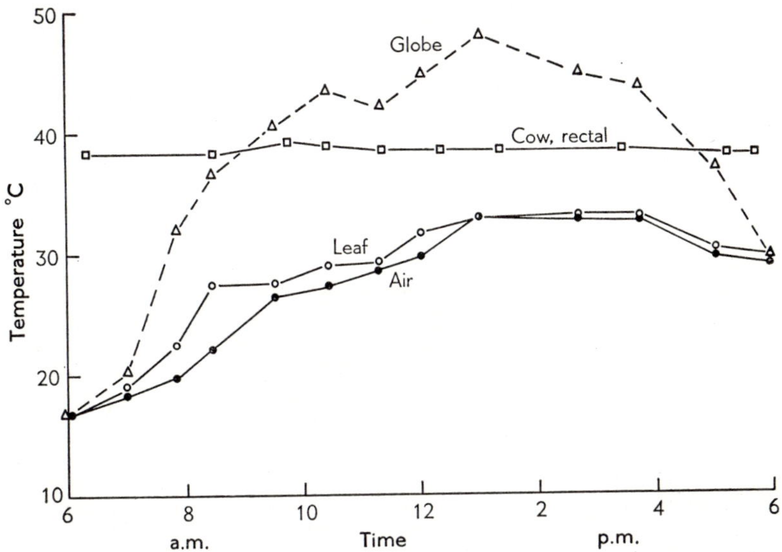

FIG. 8.2. The temperatures of 6-inch black globe thermometer, English walnut leaves, and Jersey cow, exposed to full sunlight (on 28 July 1953, at Davis, California), compared with air temperature (from Kelly, Bond & Heitman, 1954, by permission of *Ecology*).

airflow. Fig. 8.3 illustrates this difficulty by showing 'contour lines' of wind velocity over the surface of a lamb. If the flow is turbulent, there is a greater chance of approaching uniformity over the body surface, and this was the form of air movement used by Winslow and his colleagues in their work on partitional calorimetry. It is therefore necessary to consider both localized heat transfer at a given site on the animal's body and convective heat transfer from the whole animal.

Heat exchange between the animal and its respired air is a special case of forced convective heat transfer. Ingelstedt & Toremalm (1961) stress the importance of the shape of the boundary layer in determining how much heat is transferred to the inspired air.

Counter-current heat exchange in the upper airway reduces the expired air temperature to a point below that of the deep body (from 37 to about 32°C, in the case of man), and heat loss is correspondingly reduced (Walker, Wells and Merrill, 1961).

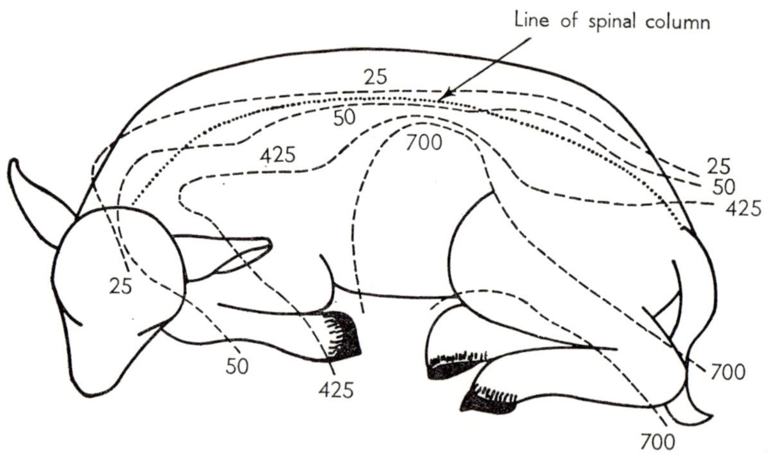

FIG. 8.3. The pattern of air movement 2 cm from a lamb, viewed from above, when exposed to a wind. The interrupted lines join points of equal velocity (cm/sec) (from Alexander, 1961*a*, by permission of *Australian Journal of Agricultural Research*).

LOCALIZED CONVECTIVE HEAT LOSS

Measurements have been made relating wind-speed to skin temperature and heat loss in a localized area on the pig's back (Mount & Ingram, 1965). In this work, pigs of three ages—1 to 2, 10 to 12 and 20 to 30 weeks—were exposed one at a time in a room the temperature of which was controlled to within $\pm 0 \cdot 5$°C, and in which radiant temperature was always within $\pm 0 \cdot 5$°C of air temperature. The wind-speeds used were close to 8, 35, 60, and 100 cm/sec (16, 69, 118, and 197 ft/min) as measured by a Simmons shielded anemometer (Tinsley and Co. Ltd, London) at a point just above the pig's surface. The air movement was produced by a speed-controlled fan. Heat loss from the selected site was measured by a Hatfield & Wilkins (1950) heat-flow disc strapped to the pig. The effects on local heat loss and body temperatures of increasing air movement are shown in Fig. 8.4.

The results obtained over a range of air temperatures from

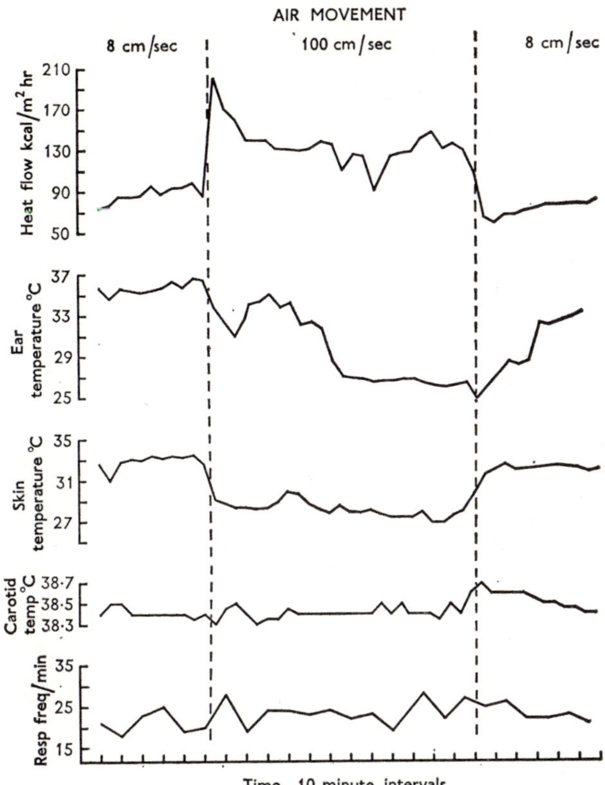

FIG. 8.4. The effects of air movement on localized heat loss from the back, body temperatures, and respiratory rate of an eleven-weeks-old pig weighing 22 kg, at an ambient temperature of 25°C (from Mount & Ingram, 1965, by permission of *Research in Veterinary Science*).

15 to 35°C are given in Fig. 8.5 and show that the increase in the rate of non-evaporative heat loss due to wind-speed was higher at the lower temperatures, even though skin temperature was lower (Fig. 8.6). The logarithmic abscissa of Fig. 8.5 emphasizes the fact that changes in wind-speed at the lower end of the scale are proportionately more effective in increasing heat loss, probably as a result of the disruption of the boundary layer which occurs at these lower speeds. Low velocity draughts of air in animal houses, for example, might be expected to influence heat loss from livestock; a simple anemometer for the measurement of low wind-speeds in such a situation has been described (Mount, 1964c). At

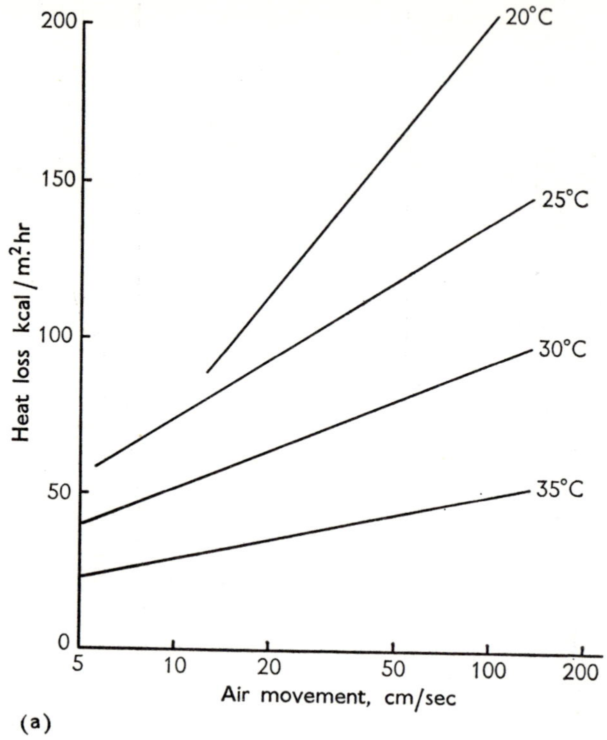

FIG. 8.5. (*a*) The effect of air movement on heat loss from the trunk in pigs aged (*a*) 1–2 weeks, 3–6 kg (from Mount & Ingram, 1965, by permission of *Research in Veterinary Science*).

higher wind-speeds convective heat loss goes on increasing, but at a decreasing rate; this is not at variance with many other measurements which have led to the general view that convective heat loss is proportional to the square root of the air velocity (Hardy, 1949; Winslow & Herrington, 1949).

Reference to Fig. 8.5 shows that the effects of increased wind-speeds at different temperatures can be equated to equivalent 'still-air' (8 cm/sec) temperatures in respect of local heat loss. Equivalent temperatures estimated in this way cannot be applied to the whole animal exposed to a wind, since the well-ventilated parts of the body would lose more heat than the less ventilated parts. The difficulty is illustrated by Kerslake's (1963) measurements of the coefficient for heat exchange by convection and

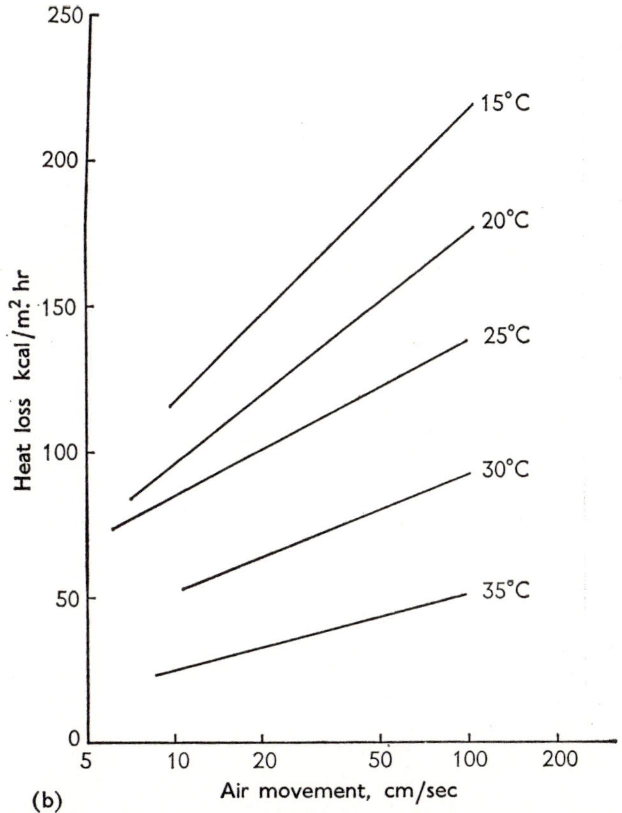

FIG. 8.5 (*b*). The effect of air movement on heat loss from the trunk in pigs aged 10–12 weeks, 20–25 kg (from Mount & Ingram, 1965, by permission of *Research in Veterinary Science*).

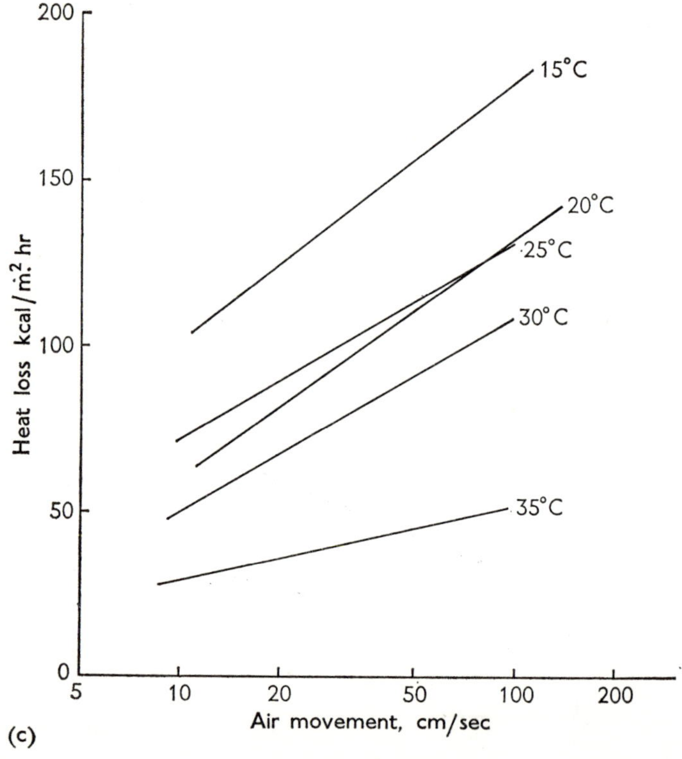

Fig. 8.5 (*c*). The effect of air movement on heat loss from the trunk in pigs aged 20–23 weeks, 60–70 kg (from Mount & Ingram, 1965, by permission of *Research in Veterinary Science*).

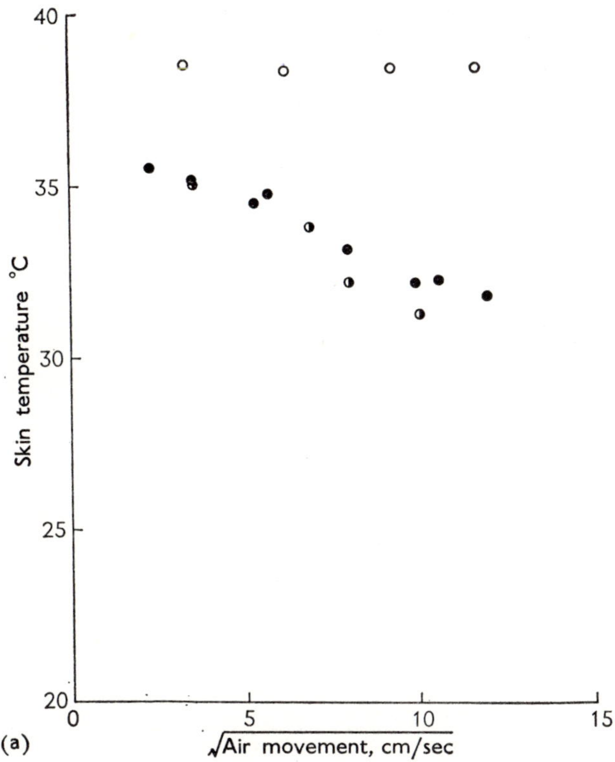

FIG. 8.6 (*a*). The effect of air movement on the skin temperature of pigs aged 1–2 weeks, 3–6 kg,

    ○ 34
    ● 25 } °C ambient temperature
    ◐ 20

(from Mount & Ingram, 1965, by permission of *Research in Veterinary Science*).

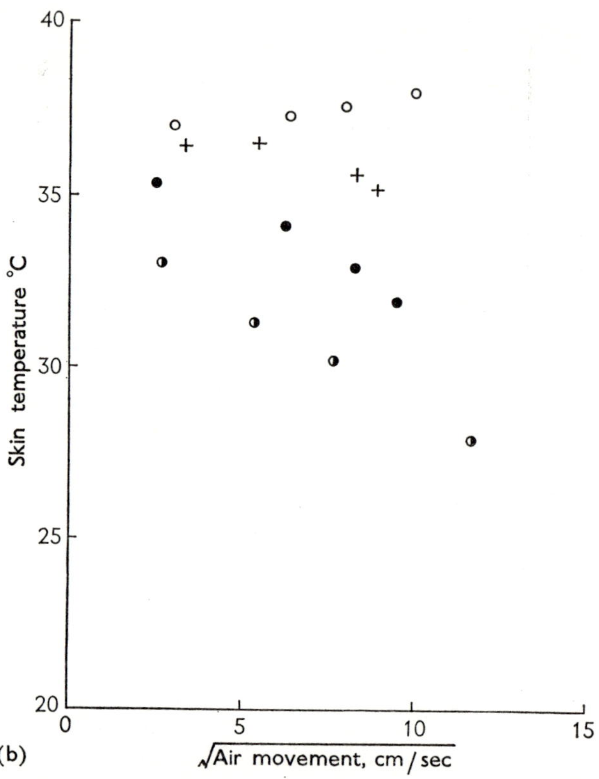

Fig. 8.6 (*b*). The effect of air movement on the skin temperature of pigs aged 10–12 weeks, 20–25 kg,

○ 34
+ 35
● 25
◑ 20
} °C ambient temperature

(from Mount & Ingram, 1965, by permission of *Research in Veterinary Science*).

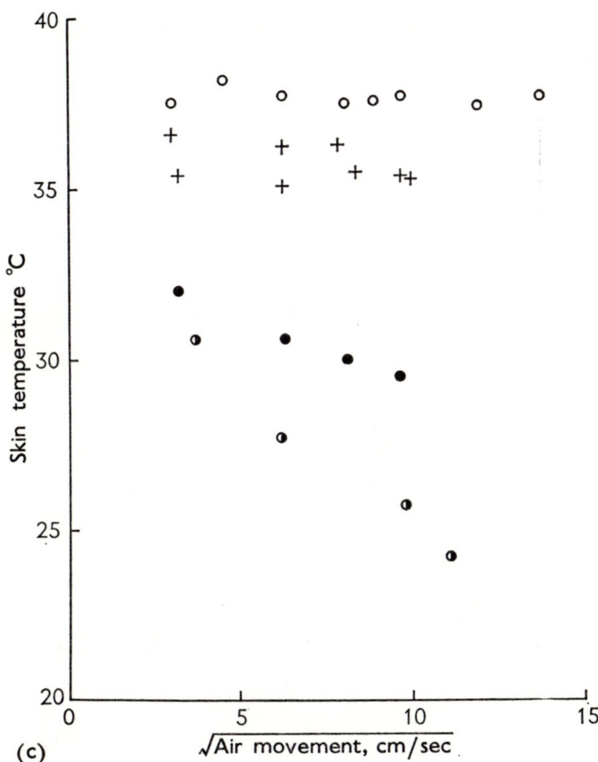

FIG. 8.6 (*c*). The effect of air movement on the skin temperature of pigs aged 20–23 weeks, 60–70 kg,

<table>
<tr><td>○</td><td>34</td><td rowspan="4">} °C ambient temperature</td></tr>
<tr><td>+</td><td>35</td></tr>
<tr><td>●</td><td>25</td></tr>
<tr><td>◑</td><td>20</td></tr>
</table>

(from Mount & Ingram, 1965, by permission of *Research in Veterinary Science*).

radiation round the perimeter of a test cylinder held in a wind. The coefficient is a measure of the heat loss per unit temperature difference between the body and the surrroundings; values of the coefficient for heat loss from the cylinder are plotted at several wind-spceds in Fig. 8.7. The general shapes of the curves at different wind-speeds are similar, with maximum heat loss at the

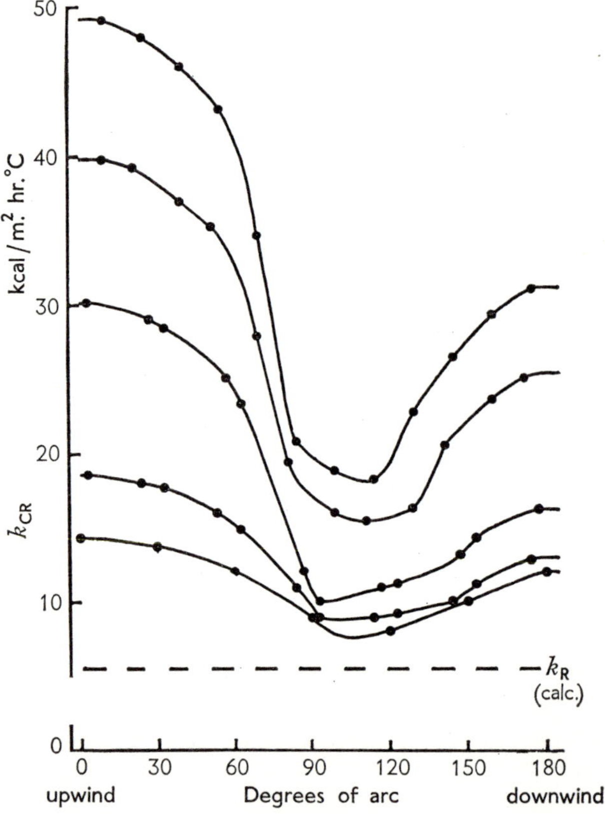

Fig. 8.7. Values of the coefficient for heat exchange by convection and radiation k_{CR} round the perimeter of an 8·9 cm diameter cylinder held vertically in a horizontal wind. Each curve shows results for a different wind speed. From above downwards the curves refer, respectively, to wind speeds of 1240, 710, 320, 120, and 50 ft/min. The dotted line shows the calculated value of the radiation coefficient k_R (from Kerslake, in *Temperature, its measurement and control in science and industry*, Volume 3, Part 3, J. D. Hardy, Editor, Reinhold Publishing Corporation, New York, 1963).

upwind surface of the cylinder, minimum at the surface where it is 90 to 100° round from the wind direction, and then a secondary peak at the downwind (180°) surface of the cylinder. Kerslake found, however, that the use of mean surface temperatures and mean heat exchange coefficients did not lead to large errors in the calculation of heat exchange, and he concluded that, provided mean values are used, it is usually unnecessary to make a detailed analysis of heat exchanges from the trunk and limbs.

EFFECT OF COAT

Tregear (1965) found from measurements on excised pelts of rabbits, horses, and pigs that while heat loss through dense fur is convectional, heat loss through sparse hair, as in the pig, is both radiative and convectional. The thermal insulation of the coat is highly dependent on the number of hairs per unit surface, and wind greatly increases heat loss from a sparsely coated skin while having a relatively smaller effect on dense fur. Tregear found the dividing line between sparse and dense in this respect to be about 1000 hairs/cm², which as he points out is rather above what many ungulates and primates have, so that these animals are highly susceptible to cooling by wind.

From measurements made by Joyce & Blaxter (1964) of the effects of wind-speed and thermal insulation in sheep with varied fleece length, it appears that at a wind-speed of 430 cm/sec (845 ft/min) a closely shorn sheep has only 45% of its thermal insulation at 27 cm/sec (53 ft/min) wind-speed. A sheep with a 30 mm fleece-length still has at the higher wind-speed 56% of its insulation at the lower wind-speed, and its lower wind-speed insulation is much higher than that of the shorn sheep (see Fig. 8.8). In spite of the destruction of coat insulation by wind, therefore, the bare-skinned animal, such as the pig, loses proportionately more heat for given wind-speed increments than a coated animal. Blaxter, Joyce & Wainman (1963) found that mild winds produced some destruction of the insulative power of the coats of cattle as well as of sheep. Lentz & Hart (1960) found the effect of air velocity on heat transfer through the fur of caribou calves to vary markedly with inclination of the surface to the direction of air movement. Water on the fur considerably increased the rate of heat transfer; at least part of this effect would be due to the displacement of highly insulating trapped air.

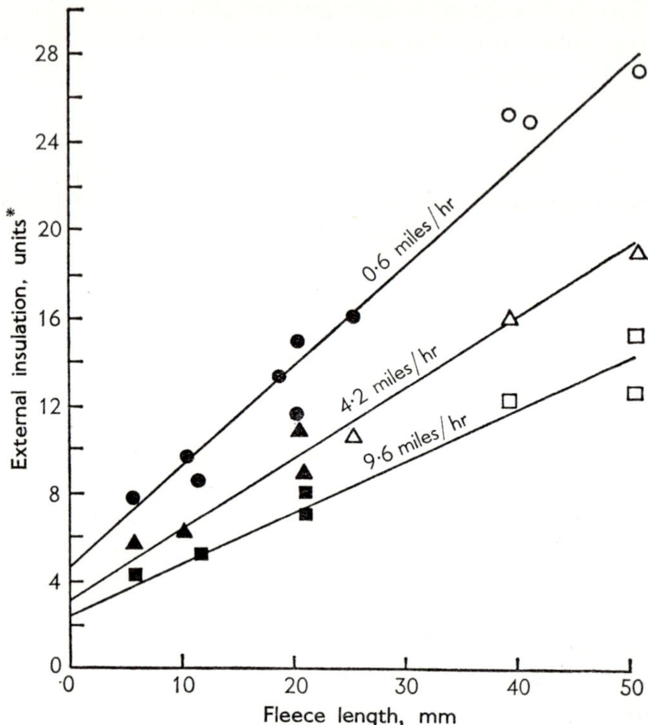

FIG. 8.8. Effect of fleece length on the external insulation of two sheep (open and black symbols) at three wind speeds. The lines are pooled within-animal regressions.

Wind speeds: ○, ●, 0·6 miles/hr;
△, ▲, 4·2 miles/hr;
□, ■, 9·6 miles/hr.

$$* \frac{\text{Temperature gradient, °C}}{\text{Heat flow (kcal) per m}^2 \text{ surface per 24 hr}} \times 10^{-3};$$

(from Joyce & Blaxter, 1964, by permission of *British Journal of Nutrition*).

CONVECTIVE HEAT LOSS FROM THE WHOLE ANIMAL

Pigs of age 1–8 days were exposed one at a time in the chamber of a respiratory metabolism apparatus to wind-speeds of 5, 34, 82, and 158 cm/sec (10, 67, 161, and 311 ft/min), estimated as mean values by an anemometer placed in the position of the pig at various points occupied by the animal's body (Mount, 1966c). Fig. 8.9 indicates that a rise in wind-speed increased the metabolic rate (which equalled the heat loss at equilibrium) at both 20 and 30°C environmental temperatures. However, as the figure shows, the increase was not as much as that found with a cylindrical model under similar conditions. Possible reasons for this are reductions in the pig's mean skin temperature (see Fig. 8.4), and reductions in the effective surface area of the pig as a result of postural change. Reductions in both skin temperature and effective

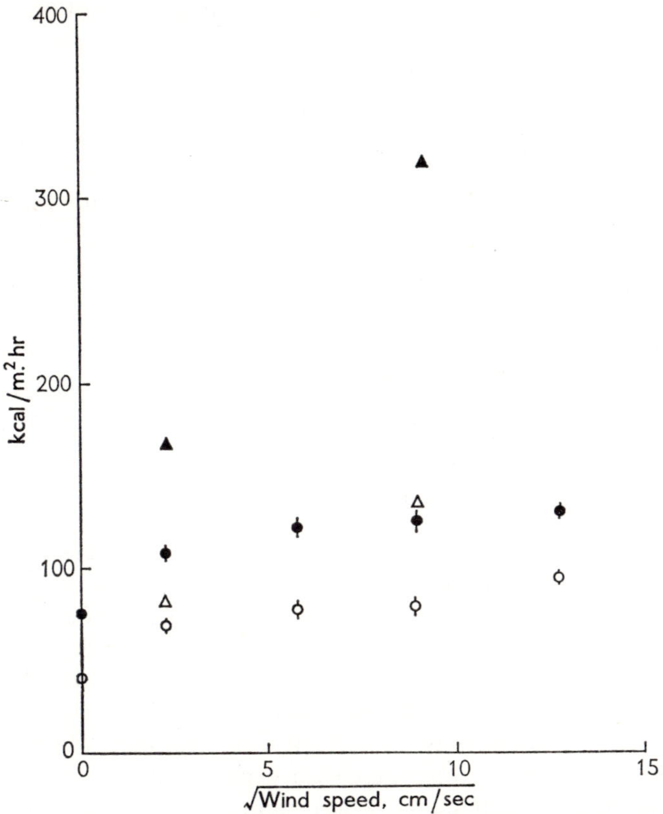

FIG. 8.9. Heat losses from pigs, and from a cylindrical model 29 cm long × 8 cm diameter, at different wind-speeds at 20 and 30°C ambient temperatures. The standard errors of the pig results are given as ordinates passing through the points.

●, pig and ▲, model at 20°C;
○, pig and △, model at 30°C

(from Mount, 1966c, by permission of *Quarterly Journal of Experimental Physiology*).

area occurred in the experiments described earlier in which the radiant environment was changed. When the effective areas for convective exchange were calculated in terms of the cylindrical model, however, impossibly low values were obtained under some conditions, suggesting the need for great care in selecting both the conditions of measurement and models for comparison with the animal. The very considerable variation in wind-speed over the animal's surface illustrated in Fig. 8.3, and the variation of the coefficient for heat exchange depending on inclination of the body to wind (see Fig. 8.7), underline the need for caution in any attempt to infer from localized measurements of heat loss how the whole animal's heat loss might vary with changes in wind-speed.

The results obtained with the young pig do not support the general statement that an animal's convective heat loss is proportional to the square root of the wind velocity. This relation may be written:

$$H_C = kA\sqrt{v}(T_1 - T_2)$$

where H_C = heat lost by convection,
k = a constant depending on size and shape,
A = surface area,
v = air velocity,
T_1 = temperature of body surface,
T_2 = environmental temperature.

Such a relation was found by Winslow and his colleagues to apply to human subjects in their calorimeter. It may be that the relative immobility and lack of postural change on the part of the subjects, and the turbulent air movement, contributed to the square root relation, which has also been found to hold for inanimate objects.

The rather unpredictable results on convective exchange contrast with measurements of radiant heat transfer, which have yielded results more consistent with expectation. For convective heat loss in still air, however, where natural convection determines the heat loss, the whole of the animal's free surface would be subject to convective exchange. The effective convective area under these conditions was found at 20°C to be 76% and at 30°C to be 86% of the total surface area (given by $m^2 = 0.097 \text{ kg}^{0.633}$), and the corresponding effective radiating areas were 67 and 76%. This is in accordance with the expectation that the convective area would exceed the radiating area, since the net radiating area is less

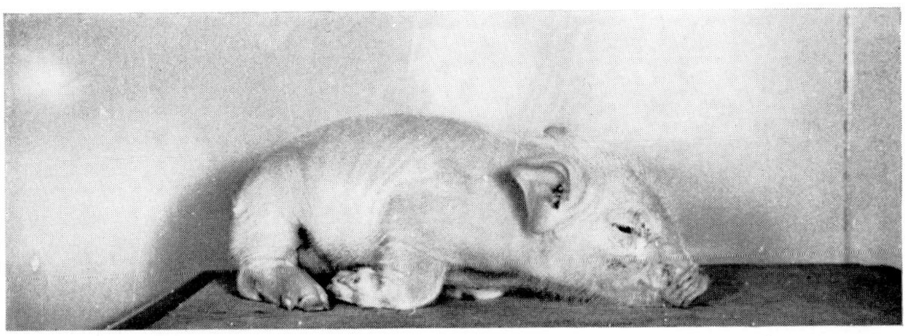

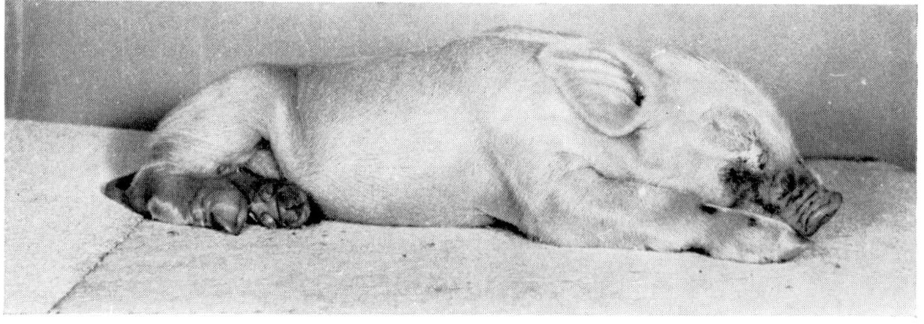

FIG. 8.11. The postures adopted by new-born pigs on cold and warm floors. In the upper photograph, at 15 °C on a concrete floor, the pig holds its thorax and abdomen out of contact with the floor. In the lower photograph, in thermal neutrality at 35 °C on a sheet of expanded polystyrene, the animal lies in a relaxed position in full contact with the floor.

than the free area by an amount dependent on the optical occlusion of one part of the body by another.

Bond, Heitman & Kelly (1965) carried out tests in a psychrometric chamber and in field production trials to determine the effects of wind on the total heat loss and skin temperatures of pigs. Wind-speeds of 18, 75, and 150 cm/sec (35, 147, and 295 ft/min) were used at air temperatures ranging from 10 to 38°C (50 to 100°F). Total heat loss increased as wind-speed increased, although the radiant component decreased due to the concomitant fall in skin temperature. The convective and evaporative components of heat loss increased.

HEAT EXCHANGE BY CONDUCTION

Compared with the quantity of work done and information gained on other channels of heat exchange, little is available on heat exchange by conduction in animals. Kelly, Bond & Garrett

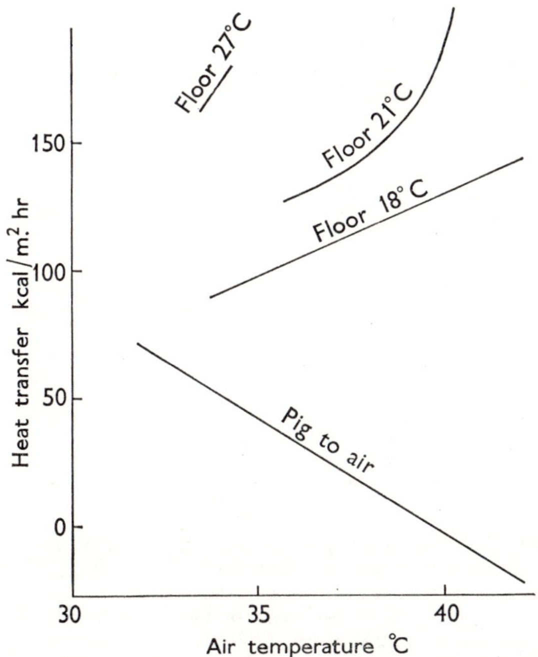

FIG. 8.10. Rates of heat transfer from pigs to floors at different air temperatures, and the rate of heat transfer to the air (from Kelly, Bond & Garrett, 1964, by permission of *Agricultural Engineering*).

G

(1964) made measurements with a heat-flow meter of the rate of heat loss from pigs to a cold slab on which the animals were lying and their results are summarized in Fig. 8.10. The lower conductive heat loss at the lower floor temperatures may be related to decreased blood flows in the skin in contact with the cooler floors. MacHardy (1960) has discussed transient heat flows from the pig to the floor, and von Pechert & Cermak (1958) investigated the relation between the growth of pigs and the heat transfer character-

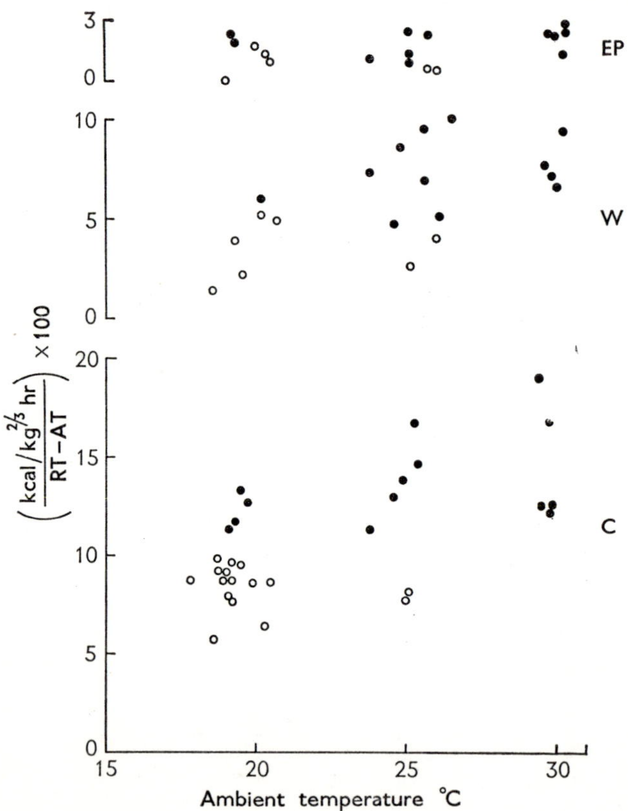

FIG. 8.12. The effect of posture on the thermal conductance between the new-born pig and the floor, over a range of ambient temperatures, on three floor materials. Thermal conductance is given on the ordinate as the ratio of heat loss to the floor over the rectal-ambient temperature difference, and therefore includes the animal's tissue insulation. Floor material: EP, expanded polystyrene; W, wood; C, concrete. Posture: ●, relaxed; ○, supported (from Mount, 1967a, by permission of *Research in Veterinary Science*).

istics of the floor. Inglis & Robertson (1953) measured heat losses
from a heated plate to different types of floor, in an attempt to find
the floor most suitable for pigs.

The actual heat loss from the new-born pig to the floor has been
estimated using a mat incorporating a gradient layer heat-flow
measurement system (Mount, 1967a). The mat rested on three
different floor materials: concrete, wood, and expanded polystyrene. The pig was put on the mat, with or without bedding
material, at ambient temperatures between 18 and 30°C. One of
the most marked effects was the decreased rate of heat loss to the

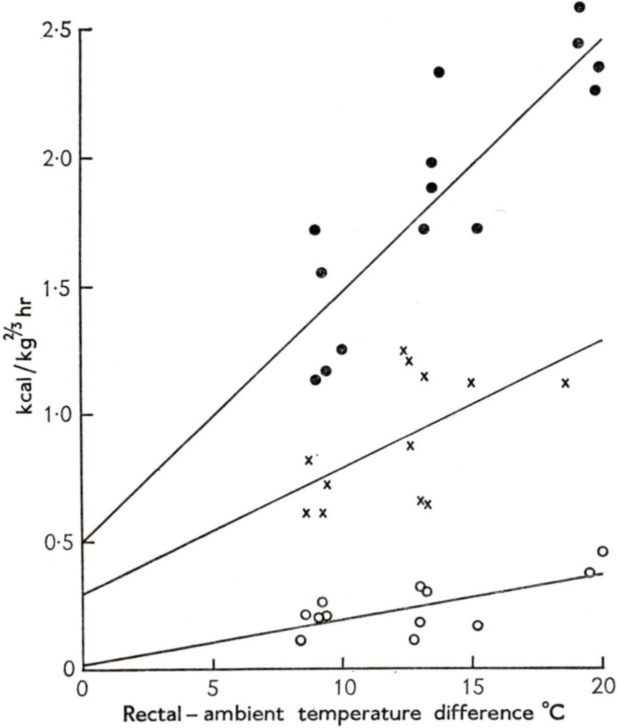

FIG. 8.13. Heat loss to the floor against the rectal-ambient temperature difference for new-born pigs in a relaxed posture on
three floor materials. Calculated regression lines:

concrete (●), $H = 0.098T + 0.50$;
wood (×), $H = 0.049T + 0.30$;
expanded polystyrene (○), $H = 0.017T + 0.023$

(from Mount, 1967a, by permission of *Research in Veterinary
Science*).

floor which resulted when the pig changed from a relaxed to a more tense posture in which it supported its trunk off the floor on its limbs (Fig. 8.11). Fig. 8.12 shows how on each type of floor the adoption of the relaxed posture increased the body-ambient thermal conductance through the floor. Fig. 8.13 gives the conductive heat-losses for a range of rectal-ambient temperature difference when the pig was in a relaxed posture; an ambient (that is, air and wall) temperature change from 30 to 20°C nearly doubled heat loss to each type of floor. At 20°C, heat loss to wood was close to heat loss to concrete at 30°C. Substituting wood for concrete was equivalent to raising floor temperature by 12°C. One inch of wood shavings on concrete reduced heat loss to the floor to 66% of that on concrete alone, and was equivalent to a rise of 9°C on a floor of bare concrete; one inch of straw reduced floor heat loss to 41% and corresponded to a 15°C rise, and one inch of woodwool reduced floor heat loss to 27% and corresponded to a 19°C rise.

The results show that about 15% of the new-born pig's total heat loss is by way of the floor when this is bare concrete; on wood it is 6%, and about 2% on expanded polystyrene. Such heat loss is not, of course, all purely conductive, although it has been termed conductive for the sake of simplicity. In the supported type of posture, particularly, heat is lost to the floor by radiation, and possibly by convection, from parts of the animal which face the floor but are not actually in contact with it. In the relaxed posture, more of the animal touches the floor and the truly conductive part of the heat flow increases. In the supported posture a smaller part of the heat loss to the floor is due to conduction. Heat loss to the floor, regardless of the mode of transmission, is the quantity which is important to the animal, and it is influenced particularly by the nature of the floor, the presence or absence of bedding, and the animal's posture.

9

THE PARTITION OF HEAT LOSS AND ASSESSMENT OF ENVIRONMENT

ON the basis of the results and discussion presented in the last three chapters, it is possible to establish the approximate partition of heat loss from the pig into the components of evaporation, radiation, convection, and conduction. The results of measurements on the new-born pig are expressed in this proportionate form in Table 9.1, for environmental temperatures of 20 and 30°C and in nearly still air. A partition which would apply over a period longer than a few hours would have to take account of the warming (or cooling) of ingested food and drink to body temperature; this is not a large fraction, amounting in the pigs kept in the Babraham pen calorimeter, for example, to about 3% of the total (Holmes & Mount, 1967).

Butchbaker & Shanklin (1964) determined radiant, convective, and evaporative heat losses from new-born pigs suspended in a nylon net in the middle of a chamber so that conductive heat loss was virtually eliminated. The chamber formed part of a gradient layer calorimeter, which also incorporated a 4π radiometer for the direct measurement of radiation. Under these conditions, radiation accounted for about 2/3 of the sensible heat loss, and convection for about 1/3, at temperatures below 32°C; evaporative heat loss was about 8% of the total heat loss at 13°C and 100% at 38°C.

Convection and radiation together usually account for the largest part of heat loss, except at high temperatures. Under certain hot conditions associated with high humidity, however, it is likely that conduction of heat to the floor is a very important factor in the animal's thermoregulation. This is probably the case in the hot humid conditions found in some pig houses, where the floor is a thin layer of uninsulated concrete resting directly on the earth. Such a floor constitutes a good 'heat sink'.

The more usual partition of heat loss from the older pig is given

TABLE 9.1

The partition of heat loss from the 2 kg pig at two ambient temperatures and for a relaxed posture, on four floors of different thermal insulation (from Mount, 1967a, by permission of *Research in Veterinary Science*)

Ambient temperature °C	Floor	Heat loss, kcal/hr						% Heat loss				
		Total	Evaporative	Non-evaporative	Radiant	Convective	Conductive	Evaporative	Non-evaporative	Radiant	Convective	Conductive
20	Expanded polystyrene and mat	15·7	1·1	14·6	7·2	7·1	0·3	7	93	46	45	2
	Wood and mat	15·7	1·1	14·6	6·8	6·8	1·0	7	93	44	43	6
	Concrete and mat	15·7	1·1	14·6	6·4	6·3	1·9	7	93	41	40	12
	Concrete alone (estimated)	15·7	1·1	14·6	6·1	6·1	2·4	7	93	39	39	15
30	Expanded polystyrene	9·5	1·2	8·3	4·8	3·4	0·1	13	87	51	35	1
	Wood and mat	9·5	1·2	8·3	4·5	3·2	0·6	13	87	48	33	6
	Concrete and mat	9·5	1·2	8·3	4·2	3·0	1·1	13	87	44	31	12
	Concrete alone (estimated)	9·5	1·2	8·3	4·1	2·8	1·4	13	87	42	30	15

in Fig. 9.1. Radiant and convective heat losses estimated for the 2 kg pig at different wind-speeds are compared with corresponding values calculated for man in Table 9.2; the partition of non-evaporative heat loss is similar in each case. Subdivisions of heat loss from the

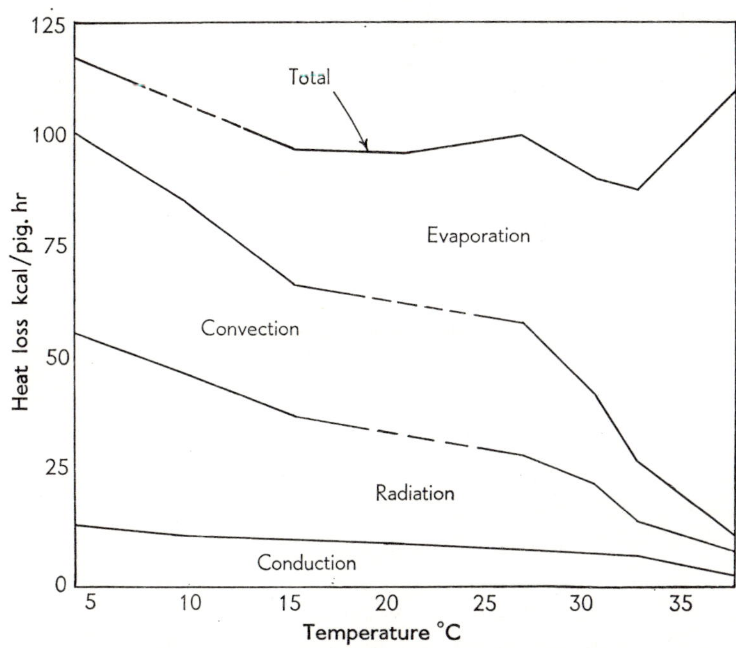

FIG. 9.1. The effect of environmental temperature on heat loss and its partition for an average pig of a group, with body weight ranging from 26 to 37 kg (from Bond, Kelly & Heitman, 1952, by permission of *Agricultural Engineering*).

TABLE 9.2

The calculated partition of non-evaporative heat loss from the 2 kg pig into its radiant and convective components as percentages at four wind-speeds at 20 and 30°C ambient temperatures. The corresponding results for man at 25°C are derived from figures given by Burton & Edholm (1955)

	Radiant			*Convective*		
Wind-speed	*Pig*	*Pig*	*Man*	*Pig*	*Pig*	*Man*
cm/sec	*20°C*	*30°C*	*25°C*	*20°C*	*30°C*	*25°C*
5	50	59	59	50	41	41
34	25	37	35	75	63	65
82	14	29	26	86	71	74
158	8	27	21	92	73	79

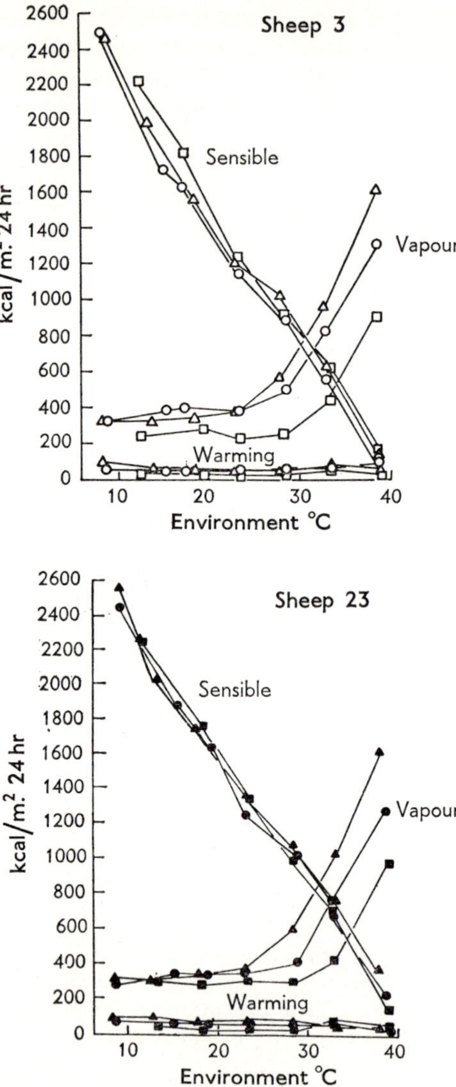

FIG. 9.2. Partition of the heat loss from two sheep into sensible (non-evaporative) and evaporative losses, and losses involved in warming food and water, per unit of surface area. Squares, low level of feeding; circles, medium level; triangles, high level (from Blaxter, Graham, Wainman & Armstrong, 1959, by permission of *Journal of Agricultural Science*).

lamb (Fig. 6.5) and sheep (Fig. 9.2) are available as evaporative and total non-evaporative losses.

THE ASSESSMENT OF THERMAL ENVIRONMENT

There have been many attempts to formulate a unitary physiological temperature scale by means of which the thermal characteristics of a given environment could be referred to as a single quantity in relation to a living organism. The demand made by the environment on an animal to produce or to dissipate heat, and so to regulate its body temperature, is formed by the effects of all the factors which influence heat exchanges through the several channels discussed in the preceding three chapters. It is tempting to think of a simple model which would integrate these components of the environment into an equivalent still-air temperature as an indicator of environmental thermal demand.

Equivalence between environments established in this way, however, as between a high-radiant–low-convective temperature combination, and a low-radiant–high-convective combination, would apply only to the particular model used, owing to the laws which govern heat transfer by radiation and convection. For example, a small sphere would be more sensitive to a given change in convective environment than would a large sphere. Coefficients could be determined to relate the results obtained with the smaller sphere to the effect on the larger sphere, but to attempt such transformations in the case of animals of differing size, thermal insulation and posture would lead to considerable complications and gross inaccuracies. It is more reasonable to abandon the search for a unitary estimate of thermal demand, and, instead, to estimate heat transfer through each of the four channels of radiation, convection, conduction and evaporation separately, and then to make a physiological assessment of the effects on the organism.

For man, however, Burton & Edholm (1955) described a method of determining an 'equivalent still-air temperature'. This is derived by adding to air temperature, algebraically, a thermal-wind-decrement in temperature, which allows for the effect of air movement. If an increment of temperature is now added to allow for radiation from the sun, or elsewhere, the 'equivalent still-shade-temperature' is estimated. These formulations have considerable convenience in assessing man's insulation requirements in different environments.

An 'effective temperature scale' was developed as a sensory scale of warmth combining air temperature, air movement, and humidity into a single index (Yaglou & Miller, 1925; Yaglou, 1927, 1949). The numerical value of the scale is the temperature of still air, saturated with water vapour, which induces a sensation of warmth or cold like that of the given condition. A 'corrected effective temperature' was also introduced which made allowance for radiant heat (Bedford, 1946).

OPERATIVE TEMPERATURE

As opposed to the sensory effective temperature scale, the operative temperature scale (Gagge, 1940) provides a physical measure of the thermal environment. Operative temperature combines as a single variable the temperature equivalents of the radiant and convective environments. This is achieved by using coefficients to relate radiant and convective heat exchanges to the differences between skin temperature on the one side, and wall and air temperatures on the other. It is possible to estimate radiant exchange with only a small error from a temperature difference, instead of from the difference between the fourth powers of the absolute temperatures (see Chapter 7), when the temperature differences are small (Burton & Edholm, 1955).

$$T_0 = \frac{K_R T_W + K_C T_A}{K_R + K_C} \tag{1}$$

where $T_0 =$ operative temperature,
$\quad K_R =$ coefficient of heat transfer by radiation,
$\quad K_C =$ coefficient of heat transfer by convection, for a given air movement rate,
$\quad T_W =$ mean radiant wall temperature,
and $\quad T_A =$ air temperature.

The following expression gives the operative temperature in terms of T_W, T_A, and V, the velocity of air movement:

$$T_0 = \frac{K_R}{K_0}\, T_W + \frac{K_C}{K_0}\left[\left(\sqrt{\frac{V}{V_0}}\right) T_A - \left\{\sqrt{\left(\frac{V}{V_0}\right)} - 1\right\} T_S\right] \tag{2}$$

where $K_0 = K_R + k_c \sqrt{V_0}$ $\tag{3}$
and $\quad K_C = k_c \sqrt{V_0}$ $\tag{4}$

K_0 = standard cooling rate, about 6 kcal/m².hr°C for man
resting in low air movement (Gagge, 1965),
V_0 = standard air movement rate.

Eq. (2) involves the skin temperature, T_S, since the equation describes the equivalent temperature in which a subject would lose the same amount of heat at a standard cooling rate, K_0, as by radiation and convection in the original environment. Operative temperature can therefore be defined in terms of an imaginary environment with uniform air and radiant temperatures with which the subject would exchange the same heat by radiation, convection, and conduction as in the actual complex environment containing the subject.

Operative temperature is thus a calorimetrically derived temperature scale, which is inclusive for the non-evaporative or sensible modes of heat transfer but which does not take evaporative

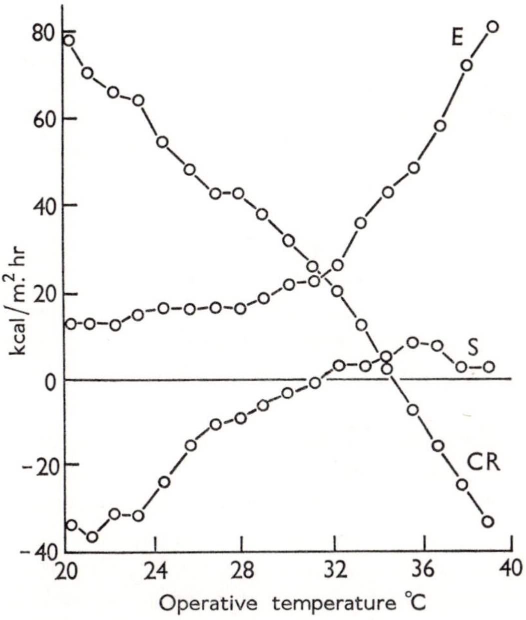

FIG. 9.3. The thermal exchanges of nude human subjects at various operative temperatures. E, heat transfer by evaporation; S, storage heat (body cooling or warming); CR, heat transfer by radiation and convection (redrawn from Gagge, Winslow & Herrington, 1938, by permission of *American Journal of Physiology*).

heat loss into account. Evaporative loss is a distinct form of regulation, and must be assessed separately. A practical disadvantage in using the scale is that the mean skin temperature must be known.

Fig. 9.3 illustrates the partition of heat loss from man over a range of operative temperature. Winslow and Herrington (1949) remark on the fact that under cool conditions (below 31°C operative temperature for naked man) the skin temperature is lower with cool air and warm walls than under reverse conditions although the operative temperature is the same on both occasions. They consider this to be due to respiratory heat loss and heat loss from body movements being governed only by air temperature.

EQUIVALENT TEMPERATURES FOR PIGS

An attempt was made by McLagan & Thomson (1950) to assess environmental conditions for pigs in terms of the Bedford scale of effective temperature. They found that a small change in effective temperature was apparently associated with a relatively large difference in pig growth, and they questioned the appropriateness of the effective temperature scale as applied to pigs. This illustrates the difficulties in applying to one organism a scale of equivalence which has been worked out for another. They found, however, that the environment which produced the most rapidly growing and apparently the healthiest pigs had the highest effective temperature.

Later work has shown the low levels of evaporative loss characteristic of the pig, and this clearly suggests that the Bedford scale, based on human sensory comparisons with a saturated atmosphere, would not be expected to be of value in work on the pig.

Although the operative temperature scale might prove to be more suitable than the effective temperature scale in application to the pig, the practical difficulties are considerable. Pigs under farm conditions live in groups, in varying physical environments and on varying levels of food intake. Under laboratory conditions it would be possible to provide standards, but only because the problem is simplified by reduction in the number of variables involved. For a practical solution, it is probably more useful at this stage to compare the maintenance energy requirements of pigs under different conditions of grouping, environment, and nutrition. Such a comparison is calorimetric, that is to say it is based on heat flow, and not on particular temperatures, whether radiant or of the air or floor. It takes into account physiological variables influencing heat loss, and the social interaction between pigs. It is an empirical approach which could be of practical use in animal husbandry (see Chapter 10).

IO

REFERENCE BASES FOR THE COMPARISON OF METABOLIC RATES

FOR purposes of calculation, it is often desirable to have metabolic rates given in terms of animals' actual body weights. For comparative purposes, however, this has considerable disadvantages. It is not useful, for example, to compare directly the metabolic rate of 4 kcal/day at 200 kcal/kg in a 20 g mouse with the 8000 kcal/day at 13 kcal/kg in a 600 kg cow, unless it is to demonstrate the effect of body size on metabolic rate per unit mass of tissue. The question which cannot be avoided in metabolic work is whether it is appropriate to calculate on the basis of the proportionality of metabolism to body weight, or whether body surface area or some other function of body size is more useful. The discussion here is concerned primarily with the homeotherm, but there is a considerable volume of work on the relation between metabolic rate and body weight in the poikilotherm. For example, Galvao, Tarasantchi & Guertzenstein (1965) investigated the heat production of tropical snakes and found metabolism to be more nearly proportional to body weight than to body surface; Cherian (1962) found the oxygen uptake of frogs to be related to the 0·925 power of body weight up to a weight of about 95 g, and to a much lower power above that weight.

Kleiber (1932) refers to the 'surface-area law' applied to homeotherms as stating that 'the basal metabolism of animals differing in size is nearly proportional to their respective body surfaces'. This rule was first put forward by Sarrus & Rameaux in 1839, and for many years subsequently workers in the metabolic field tended to regard their results as being at fault if they did not show proportionality to surface area. Brody (1945) and Kleiber (1961) have discussed this subject at length. Table 10.1 gives results from various animals in terms both of body weight and of surface area. When compared with body weight, surface area offers a much more uniform basis for comparison between species.

TABLE 10.1

Voit's Table on the Surface Law of metabolism (from Kleiber, 1961, by permission of John Wiley & Sons, Inc.)

Animal	No. of determinations	Average weight, kg	Fasting metabolism kcal per day	
			per kg	*per m²*
Horse	8	441	11·3	948
Pig	2	128	19·1	1,078
Man	5	64·3	32·1	1,042
Dog	15	15·2	51·5	1,039
Rabbit	5	2·3	75·1	776
Goose	6	3·5	66·7	1,018
Hen	2	2·0	71·0	1,008

It has been remarked that subtraction of the ear surface area of the rabbit brings the rabbit figure of 776 up to 917, which makes it conform more nearly with the requirement of the Surface Law!

It is necessary to find some common basis for comparing the metabolism of animals different in size, so that the effect on metabolism of influences other than body size can be studied. As Kleiber (1961) points out, this is a rational objective, and it is curious that Benedict (1938), who measured metabolic rate over such a wide range of animals, should not have applied his extensive results in pursuit of this aim; instead, Benedict rejected this approach. The relation of metabolic rate to body size has a highly practical corollary in its application to the determination of feeding standards for the maintenance requirements of farm animals. The relation is affected by factors in the environment, the animal and its nutrition; the practical problem lies in how the relevant factors can be taken into account in as simple and yet as useful an expression of metabolic body size as possible.

Three approaches to the problem are by (1) calculation of heat transfer, (2) determination of active body mass, and (3) using simple relations between metabolic rate and body weight. These will be considered in turn.

HEAT TRANSFER

Heat transfer depends on thermal insulation, both the specific insulation (per unit of surface area of the organism) and the overall insulation (depending on body size, shape, posture, and movement). The reciprocal of insulation is conductance; the closest simple approach to an index of thermal conductance, to which heat transfer would be proportional, is the total surface area, which has been extensively used as a metabolic reference base (Brody, 1945; Kleiber, 1961). However, this is not the effective surface area for heat exchange by radiation, convection, and conduction; the actual areas for exchange are affected by posture (see Chapters 7 and 8). In addition, specific insulation varies between species, between animals, and on different parts of the same animal, so that surface area by itself cannot provide a generalized basis for metabolic rate comparisons between different animals even though environmental conditions may be standardized. What is needed instead is an expression which takes size, shape, postural variation, and specific insulation into account for given ranges of environmental conditions.

In the steady state, an animal's heat production is equal to its heat loss. The total non-evaporative heat flow is then related to the body-environment temperature gradient by:

$$H_\mathrm{S} = \frac{T_\mathrm{R} - T_\mathrm{E}}{R} \tag{1}$$

where H_S = total non-evaporative heat flow, kcal/hr
T_R = deep-body (rectal) temperature, °C
T_E = environmental (air and radiant, or operative) temperature °C
R = *total* thermal insulation between the animal's core and the environment; it includes both the internal (or tissue) and the external insulations, °C hr/kcal.

The reciprocal of the insulation is the thermal conductance, which depends on the same factors as those affecting insulation, but in the inverse sense. The thermal conductance is proportional to

1. Some function of air movement, which includes the effects of the animal's movements in causing air to flow over its surfaces;
2. a function of posture; in the more extended position an animal loses more heat than it does when the limbs and body are flexed; in addition, parts of small radius—limbs and digits—are exposed during extension, and these parts lose more heat per unit area for a given temperature gradient than do parts of larger radius (Hardy, 1949);

3. some function of body size, $f(B)$;
4. the reciprocal of the mean *specific* insulation.

For convenience, the effects of air movement and posture can be considered together, since increasing air movement and increasing postural extension both lead to increased heat flow. If these are combined in a 'form factor' F, then:

$$\text{total conductance} = \frac{F.f(B)}{R_S} \tag{2}$$

where $\quad R_S$ = mean specific insulation, °C.m².hr/kcal, where m² refers to the total body surface area.

$$\text{Therefore} \quad R = \frac{R_S}{F.f(B)} \tag{3}$$

Substituting in eqn. (1):

$$H_S = \frac{F.f(B)(T_R - T_E)}{R_S} \tag{4}$$

If F is to be a dimensionless number, then the dimensions of $f(B)$ are

$$\frac{°C.m^2.hr.\ kcal}{kcal.\ °C.\ hr} = m^2$$

so that the function of body size which is required is the surface area, A.

Eqn. (4) then becomes

$$H_S = \frac{FA(T_R - T_E)}{R_S} \tag{5}$$

A great deal of effort has been put into the determination of surface areas in different animals, using roller-integrator, photographic, and other methods (Brody, 1945; Kleiber, 1961). The value of A in eqn. (5) can be determined from the appropriate formula. For the pig (Brody, 1945),

$$A = 0.097 W^{0.633} \tag{6}$$

where $\quad A$ = surface area, m²
$\qquad W$ = body weight, kg.

This equation was used by Brody for pigs weighing up to 250 kg.

Both the form factor (F) and the mean specific insulation (R_S) contribute to the total thermal conductance between the animal and its environment. If R_S could be determined, it would be possible to calculate a range of values of F for different conditions of posture and air movement. The specific insulation, however, varies over the animal's surface, being lower on the limbs than on the trunk; this is markedly so for a coated species such as the sheep. The mean value of the R_S elements which actually affect heat exchange depends on posture, the exposure of poorly insulated limbs, activity, and air movement, and is in this way affected by the same variables which influence the form factor (F). Kleiber (1932) presents calculations of specific insulation for rabbits, steers, and sheep showing the markedly lower values at high environmental temperatures, and, in the case of rabbits, the effect of age in diminishing the temperature-dependence of the specific insulation in older rabbits. In practical terms it is therefore more useful to combine F and R_S as F/R_S which gives an overall thermal conductance factor, C. From eqn. (5):

$$C = \frac{H_S}{A(T_R - T_E)} \tag{7}$$

The mean slopes of the plots of H against $A(T_R - T_E)$ in Fig. 4.6 give values of C for single pigs from birth to about five weeks of age. For pigs less than one week old, both 5°C and 30°C are below the critical temperature, so that peripheral vasoconstriction would be expected in both cases. The difference in conductance-value between the two temperatures probably reflects a postural difference, with parts of smaller radius and lower R_S being exposed at 30°C and occluded by other parts of the body at 5°C. Measurements of effective radiating area support this conclusion (Mount, 1964*b*). In addition to air temperature, another factor affecting posture, and therefore the conductance-value, is the nature of the floor on which the pig lies and the rate of heat loss to the floor (Mount, 1967*a* and see Chapter 8). On a 'colder' floor the animal adopts a more flexed, tense posture.

Apart from factors in the physical environment, factors connected with the animal also affect the result. The metabolic rate of new-born pigs is related to the deep body temperature under conditions of maximum and minimum metabolism. The grouping of animals together reduces heat losses in the cold, by modifying the individual animal's thermal environment. This effect occurs both in litters of new-born pigs and in older pigs (see Figs. 5.6, 5.7, and 5.9).

The conductance-value method of formulating metabolic body size is thus a return to a form of the surface-area law. Such a law suffers from being too much of a generalization, which means that in the individual case too much error is often involved. The incorporation of terms for temperature gradient and total conductance in eqn. (7) allows for a more specific application of surface area in predicting heat transfer in given conditions.

In practice, however, the animal adjusts to a changing environment in such a way that as $(T_R - T_E)$ in eqn. (7) becomes larger, so C becomes smaller, through effects on both the specific and overall insulations. The result is that the non-evaporative heat loss H_S tends to remain constant; as the temperature falls progressively, H_S does, of course, rise, but not as much as if C were to remain constant. As a modification of the surface area law of metabolism, therefore eqn. (7) is of limited usefulness since the surface area, A, has to be multiplied by two inversely related quantities, C (the overall thermal conductance factor) and $(T_R - T_E)$, to give the non-evaporative heat flow, H_S. At temperatures close to thermal neutrality, the product $C(T_R - T_E)$ tends towards a constant value, and the equation reduces to a statement of the surface area law, $H_S/A = \text{constant}$.

ACTIVE BODY MASS

When metabolic rate is either at a maximum or at a minimum, the rate is an expression of the upper or lower limits of the organism's range of potential metabolic activity. This is because at these two extremes no regulation of rate is required, so that the upper and lower limits can be measured. The limits are a measure of the total metabolic activity of all the tissues of the organism when the environmental conditions are such that they demand either the maximum heat production of which the organism is capable, or the minimum. For this reason, the limits of activity represent an integration of the several products of tissue mass and metabolic rate per unit mass for the different tissues of which the organism is composed. The mass of a given tissue depends on the organism's history of activity, environmental exposure and nutrition (Barnett & Mount, 1967). For example, the somatic muscles of an athlete constitute a greater part of the body than in the case of an office worker; rats exposed to winter cold have organ weights and proportions different from those of rats kept in the laboratory under equable conditions; an under-fed animal has less fat and muscle than a well-fed animal. The metabolic rate per given mass of tissue is similarly dependent on other factors. It depends partly on the size of the animal, being higher in a smaller than in a larger animal. It also depends on the history of acclimatization to environment, type of food, and the level of activity demanded of the tissue; thus non-shivering thermogenesis is much more highly developed in cold-exposed rats than in rats kept in the warm (Davis, Johnston, Bell & Cremer, 1960).

A measurement of 'lean body mass' can therefore provide a satisfactory metabolic reference base only if the lean mass estimate itself reflects not only the mass of tissue but also the level of enzyme activity in the organism. Although this raises unanswered questions, there are other reasons for not adopting the active body mass (a concept first introduced by Rubner in 1902) as a reference base. While such a reference base may be satisfactory for animals similar in age, weight, and state of adaptation, and under standard conditions, particularly near the basal level of metabolism, it is inapplicable to other conditions because the metabolic rate is then not determined by the potential metabolic activity for maximum or minimum, but is instead set by the interaction of organism and environment, operating through neuro-humoral control mechanisms, at some point between the maximum and minimum. The controlling factor is then not the body mass, but the rate of heat transfer between organism and environment.

Active body mass is therefore not satisfactory as a basis for the general comparison of metabolic rates between animals, although it may suffice for limited comparisons under given conditions.

METABOLIC RATE AND BODY WEIGHT

From large numbers of measurements of metabolic rates of animals of different weights it is clear that the logarithm of the metabolism tends to vary directly with the logarithm of the body weight, according to the expression:

$$\log M = b \log W + \log a$$

where M = metabolic rate,

          W = body weight,

and a and b are constants.

This is equivalent to:

$$M = aW^b$$

The double-log plot provides a linear relation, and so allows the exponent b to be calculated from the slope, which is given by the regression coefficient. W^b is then termed the metabolic body size.

A large number of different values has been determined for b under different experimental conditions. Alternatively, b can be fixed, and the quantity a varied to fit the results. One example, where $b=1$, makes metabolic rate proportional to body weight; b may also be fixed in relation to some other parameter, such as surface area. In the special case of the surface areas of bodies similar in shape and density the exponent then becomes 0·67, as in the Meeh (1879) surface area formula:

$$A = kW^{0.67}$$

where $A =$ surface area, and k is a constant.

Brody (1945) and Kleiber (1932, 1947, 1961, 1965) have discussed the exponent at length. Brody found an interspecific value of 0·734 to fit a wide range of values obtained by a number of investigators; he subsequently approximated this, first to 0·73, and then 0·7. Kleiber, however, found a value of 0·74 as the best regression coefficient, and rounded this to 0·75 as a convenient value for calculation. Kleiber disapproved of Brody's figure of 0·7, since this is too close to the two-thirds power; Kleiber considered a three-quarters power rule to be more appropriate to the relation between metabolic rate and body weight. Brody's 0·73 was the value adopted by a National Research Council (U.S.) Conference in 1935, in preference to 0·75. Such fine discrimination is, however, questionable in view of the considerable variation in the measurements on which the exponent is based. Kleiber (1965) commented that a significant difference between proportionality to the two-thirds power of body weight and the three-quarters power could be established only with groups of animals showing a weight range of ninefold or more in extent; to establish a significant difference between proportionality to body weight or to the three-quarters power requires a threefold weight difference (Fig. 10.1). At a meeting of the European Association for Animal Production, Energy Metabolism Symposium, in 1964, the three-quarters power of body weight was adopted as the *interspecific* reference base. The comparison between the metabolic rates of piglets and human infants in Chapter 4 is made on the three-quarters power basis.

Some approximation of this sort is convenient for comparison between animals of different species. Kleiber has arrived at 70 kcal/kg$^{0.75}$ per day as the mean value for standard metabolism in different species. This holds reasonably well, at least as a fair guide, in the older animal. Kleiber's actual mean value was

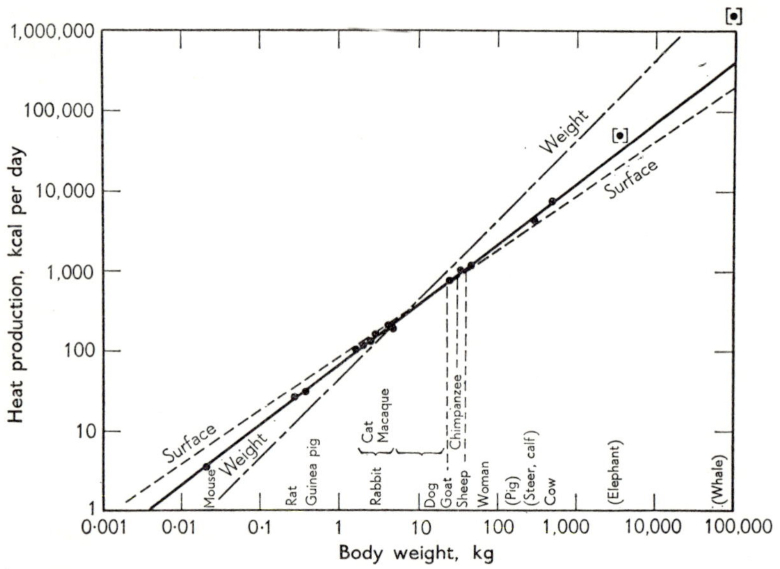

FIG. 10.1. Relation of the logarithms of metabolic rate (*M*) and body weight (*W*) for a number of mammals. The regression coefficient is 0.756 ±0.004, marked on the diagram as the continuous line. The interrupted lines indicate the slopes required for proportionality to weight or to surface area (from Kleiber, 1961, by permission of John Wiley & Sons, Inc).

69.5 kcal/kg$^{0.75}$ per day, while Brody (1945) found 70.5 kcal/kg$^{0.73}$ per day. In respect of the growing pig living in a group in a pen, close to optimal conditions, however, the resting value is closer to 140 kcal/kg$^{0.75}$ per day for 20 kg pigs and 135 for 60 kg pigs. The corresponding mean 24-hour values are approximately 160 and 150 kcal/kg$^{0.75}$ per hour. Fasting would lower this value. Blaxter (1962*b*) found the fasting metabolism of wether sheep aged more than two years, ranging in weight from 27 to 72 kg, to be about 58·5 kcal/kg$^{0.73}$, which is 20% lower than the interspecies mean. With cattle, however, Blaxter & Wainman (1966) found a fasting metabolism of 100 kcal/kg$^{0.73}$ per day in Ayrshire steers, 81 in Black cattle of the Aberdeen Angus breed, and 96 in Ayrshire × Beef Shorthorn cattle. From their studies on feeding standards for dairy cattle, Hashizume, Morimoto, Hamada, Masubuchi, Abe, Horii, Tanaka, Zitsukawa, Yokota & Anbo (1964) found the heat production in kcal per kg$^{0.75}$ per day to be 119 at 5°C, 117 at 17°C, and 113 at 27°C.

Hemmingsen (1960) has discussed the subject of energy metabolism and body size at some length, ranging from unicellular organisms to large mammals, and even to beech trees. In general he found, from interspecific calculations involving considerable ranges of weight, that metabolism is related to weight raised to the 0·75 power for poikilotherms as well as homeotherms; the slope of the three lines in the log-log plot of Fig. 10.2 is 0·75 in each

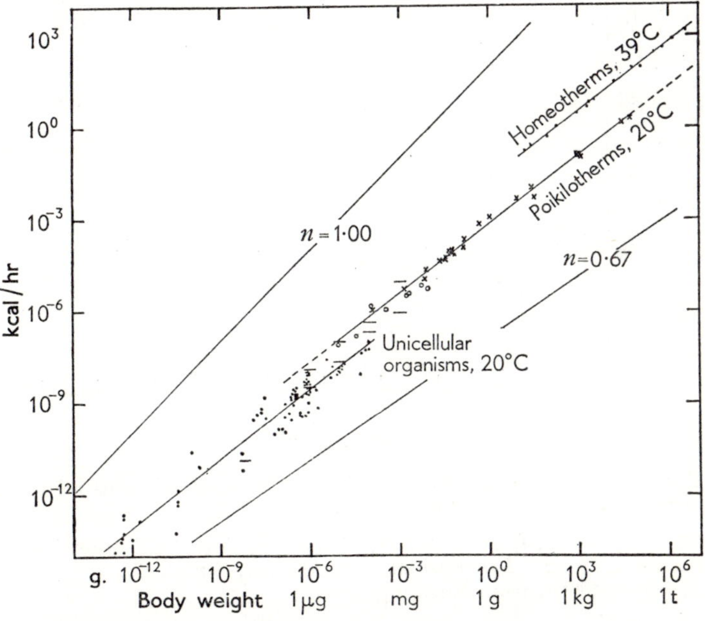

Fig. 10.2. Double logarithmic plots of metabolism against body weight for unicellular organisms and larger poikilotherms at 20°C, and for homeotherms. The two outer lines are drawn to demonstrate proportionality of metabolism to body weight ($n = 1·00$) and to body surface area ($n = 0·67$), (from Hemmingsen, 1960, by permission of the author).

case, a smoothed value obtained by the method of the least sum of squares assuming the exponent to be the same for each of the three sets of values. Hemmingsen also found the maximal metabolism, between species, to be related to the three-quarters power of body weight.

Age

It would be expected that factors in the animal and environment would influence the exponent in a particular case, just as they do when metabolic body size is sought by calculations on the basis of

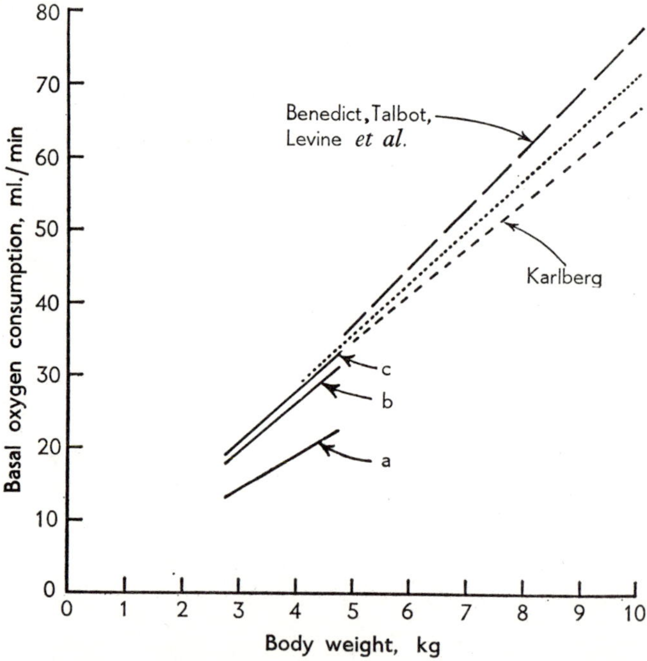

FIG. 10.3. Basal oxygen consumption (M) of human infants related to body weight (W) on linear scales.

$M = 6 \cdot 35 W^{1 \cdot 090}$, from Benedict, Talbot, Levine *et al.*, quoted by Karlberg (1952) *Acta Paediat.*, Stockh., **41**, Suppl. 89, 11–151; $M = 8 \cdot 14 W^{0 \cdot 918}$, from Karlberg (1952); $M = 7 \cdot 2 W$.

a, 0–6 hr of age, $M = 4 \cdot 76 W$;
b, 18–30 hr of age, $M = 6 \cdot 59 W$;
c, 6–10 days of age, $M = 7 \cdot 02 W$

(from Hill & Rahimtulla, 1965, by permission of *Journal of Physiology*).

heat transfer. Mount & Rowell (1960*a*, *b*) examined the relation between metabolic rate and body weight, rectal temperature, and age, in pigs during the first five weeks after birth (see Chapter 4). Multivariate regression analysis was used so that the effects of each variable could be examined in turn. In this way it was found that the exponent of body weight was 0·61 for first-week pigs at 30°C environmental temperature, 0·56 for first-week pigs at 4°C, and 0·68 for pigs aged 1–5 weeks at 30°C. These were the values obtained by calculations with rectal temperature and age held

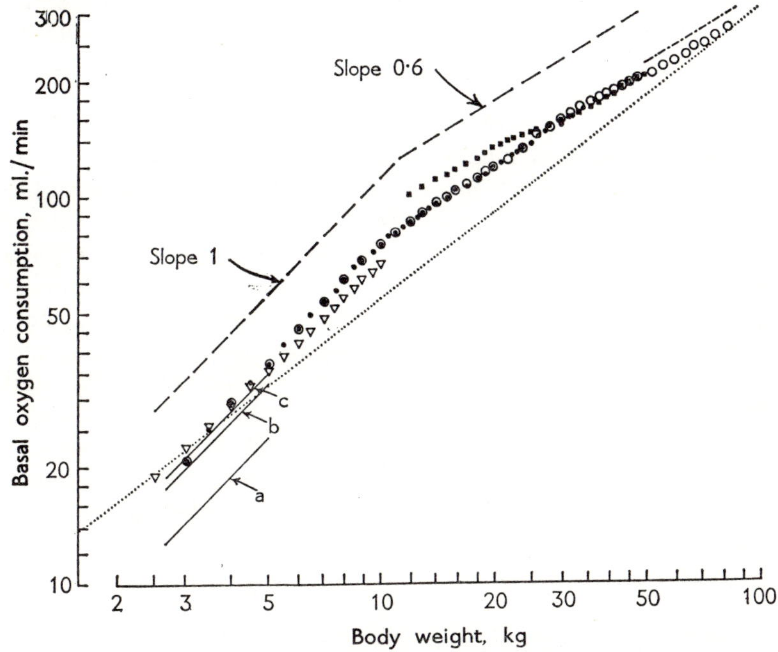

FIG. 10.4. Basal oxygen consumption of man related to body weight on double logarithmic co-ordinates, covering the entire period from birth to adult life.

●, Benedict & Talbot (1921) Carnegie Inst. Publ. 302;
○, Talbot (1938) *Am. J. Dis. Child.* **55,** 455;
■, Lewis *et al.* (1943) *J. Pediat.* **23,** 1;
▽, Karlberg (1952) (see legend Fig. 10.3);
—·— von Döbeln (1956) *Acta physiol. scand.* **37,** suppl. 126, 7.

Lines a, b, and c as in Fig. 10.3. Slopes of 1·0 and 0·6 drawn in interrupted lines.
. . . . . . from Kleiber (1947), kcal/24 hr $= 70W^{0.75}$
(from Hill & Rahimtulla, 1965, by permission of *Journal of Physiology*).

constant. As a mean result for the first week pigs at both 4 and 30°C, however, an exponent of 0·60 was found when the effects of varying rectal temperature and age were removed in this way, while an exponent of 0·86 resulted when these two variables were ignored. It is clear, therefore, that under the given conditions of *intra-specific* measurements caution must be exercised in attaching any undue significance to particular exponents.

The relation between basal metabolic rate and body weight in the human infant (Hill & Rahimtulla, 1965) shows that the rate per kg remains almost constant during the first year after birth, indicating an exponent of about unity (Fig. 10.3). Later in life the exponent decreases to 0·6; this is illustrated in Fig. 10.4, in which Kleiber's mean standard metabolic rate of $70 \times (\text{weight in kg})^{0.75}$ is plotted for comparison. The plotted points refer to *basal* metabolism. It is of interest that the young human, particularly between weights of 10 and 20 kg, but also extending on either side of that range, shows basal rates in excess of Kleiber's predicted standard rate. This is in accord with the higher rates found in the growing pig. Kleiber's three-quarters power law would appear to be more appropriate to the mature organism than to the young one.

Brody (1945) refers to variation in the exponent in young rats during the early growth period. When the logarithm of total metabolism is plotted against the logarithm of body weight, there are three breaks in the plot: at about ten days of age, at weaning at about three weeks, and at about six weeks. The first three segments of the curve have slopes varying from 0·8 to 1·1, whereas the fourth segment has a slope of only about 0·2.

Pigs

From continuous 24-hour measurements of heat loss from groups of growing pigs living in a calorimeter designed as a pig pen, Holmes & Mount (1967) found that the exponent which gave proportionality of heat loss to body weight varied with age and probably with the plane of nutrition. The regression equations for log heat loss on log body weight at 20°C environmental temperature are given in Table 10.2, showing values ranging from 0·4 to 1·0. 0·4 applies to 60 kg pigs on a constant restricted food intake, while 1·0 applies to younger, 20 kg pigs, receiving a food intake which rises in direct proportion to body weight. In Table 10.3 these results are presented with others to form an ascending order of exponents of body weight. In addition, Brody & Kibler (1944)

TABLE 10.2

Regression equations for logarithm of heat loss from groups of pigs during 24-hour day, against logarithm of body weight, at 20°C. The 20 kg pigs received food intake proportional to body weight; the food intake of the 60 kg pigs was restricted at a constant level. $y =$ log heat loss (kcal/pig hour); $x =$ log body weight (kg) (from Holmes & Mount, 1967)

Approximate weight range, kg	Regression equation
20–60	$y = 0.64(\pm 0.02)x + 0.98$
20–60	$y = 0.81(\pm 0.02)x + 0.69$
20	$y = 1.00(\pm 0.07)x + 0.55$
60	$y = 0.42(\pm 0.12)x + 1.40$
20–60	$y = 0.65(\pm 0.02)x + 0.99$
20–60	$y = 0.73(\pm 0.02)x + 0.84$
20–60	$y = 0.74(\pm 0.03)x + 0.81$

TABLE 10.3

The range of approximate body weight exponents which confer proportionality to heat loss in pigs at different body weights and levels of food intake. In the expression $M = aW^b$, b is the exponent, M is the heat loss, W is the body weight, and a is a constant.

Food intake	Body weight, kg	Exponent	Source
Constant	60 to 72	0.4	Holmes & Mount, 1967
$\propto W^{0.7}$	20 to 100	0.6	Ludvigsen & Thorbek, 1955
$\propto W^{1.0}$	34 to 64	0.8	Holmes & Mount, 1967
ad libitum	20 to 45	0.9	J. D. Pullar, personal communication
$\propto W^{1.0}$	17 to 34	1.0	Holmes & Mount, 1967
ab libitum	5 to 12	1.1	Cairnie, 1958

found that heat loss increased proportionally to (body weight)[0.9] in pigs up to 60 kg body weight, with exponents ranging from 0.3 to 0.8 in heavier pigs.

The indications are that in pigs weighing 20–40 kg, the exponent is close to unity if the food intake increases rapidly as the animal grows. If food intake increases less quickly than body weight, the exponent is less than 1.0. The older the pigs, and the lower the plane of nutrition, the smaller is the exponent. It is quite clear that no single value of the exponent can be applied over this range of conditions, in spite of the limitation of environmental variation within a roughly optimal zone, using the term optimal to imply minimum maintenance energy requirements of the animals with correspondingly high utilization rates of food for growth.

THE USEFULNESS OF METABOLIC BODY SIZE

The brief outlines of the three approaches made above to the formulation of metabolic body size in the pig suggest that the adoption of a rigid formula, such as surface area or a fixed exponent of body weight, would be inadequate and would lead to considerable error on many occasions, while the concept of active body mass is open to many objections. Variations in the animal, its physical environment and its level of food intake affect both the conductance-value, which is derived from heat transfer considerations, and the exponent of body weight, which is derived from the log-log relation between weight and metabolism.

It is, however, possible to determine ranges of either conductance-value or body weight exponent for pigs kept under given conditions, if the important controlling factors can be identified. If the physical environment is close to providing the conditions of thermal neutrality, the important factors are the age and size of the pig and the plane of nutrition. Provision of thermal neutrality implies that such factors as the size of the group, presence of bedding, and the animals' thermal insulation, are taken into account already; these additional factors, however, could be recognized explicitly as variables so that formulation of body size could be made over a range of conditions. As between conductance-value and exponent, the latter is probably the more useful, since it is simpler; the conductance-value has no compensating counter-advantage. Values of the body-weight exponent could be provided in table form; in this respect, Table 10.3 provides a beginning, which can be amplified and elaborated to any degree consistent with the limitations of error inherent in the method, which are due largely to between-animal variation.

This conclusion is in keeping with that arrived at by Gridgeman and Héroux (1965), who recommended that the relation between metabolic rate and body weight should be determined on a within-experiment basis. The great advantage of this approach is that it does not involve any assumptions about a definite form of generalized relation between the two quantities applying to animals of either one or several species. The metabolic rate and body weight are instead related together under the conditions of environment and nutrition obtaining at the time of measurement. The approach does not provide a general prediction of metabolic rate in animals, but in any case what seems to have emerged is that there is no precise prediction method, at least not precise enough, for example, to act as a general indicator of maintenance feeding standards. An approximate guide is provided by calculating metabolic rate as proportional to either the two-thirds or three-quarters power of body weight, multiplied by a suitable factor. Such a factor to multiply (weight in kg)$^{3/4}$ would, for growing pigs, lie between 135 and 160, to give kilocalories per 24 hr.

Thus in attempts to arrive at a suitable reference base for the comparison of metabolic rates of different animals, both rational and empirical approaches have been made to the problem. The rational approach breaks down because many variables are not measurable in practice. One is therefore thrown back on the empirical approach, which has its own limitations.

11

NEUROLOGY AND BEHAVIOUR OF THE PIG
IN RELATION TO ENVIRONMENT

IT is not possible to give a comprehensive account of behavioural responses in the pig since the study of the animal has not been sufficiently extensive. The work which has been done has originated primarily from an interest in pig husbandry and the use of the animal as a producer of meat for human consumption.

Hafez, Sumption & Jakway (1962) have given an account of the behaviour of pigs, and Fox (1967) has discussed the influence of domestication upon the behaviour of animals. The pig's adaptability is indicated by the rapidity with which wild pigs become accustomed to restricted conditions. It is probable that the popular concept of the pig as a dirty and greedy animal has led to false conclusions with regard to its behaviour patterns, since to a large degree the dirt and greed have been consequent on human control of the animal.

The available information can none the less be pieced together to provide at least some basis on which the animal's behaviour can be understood. There has been little definitive investigation into the nervous system, but rather more information is available on the behaviour of pigs.

NEUROLOGY

Motor cortex

The motor cortex of the anaesthetized pig has been studied with electrical stimuli applied to the cortical surface and observation made of consequent muscular movement (Breazile, Swafford & Thompson, 1966). Somatic motor activity was elicited from a limited area in the superior longitudinal gyrus, between the coronal and cruciate sulci (Fig. 11.1). From rostral to caudal, this region represented ipsilateral upper lip, ipsilateral lower lip, contralateral upper lip, contralateral lower lip, fore limb, hind limb. Fore limb and hind limb muscle activity was elicited by stimulation of the medial aspect of the superior longitudinal gyrus, with no representation on the crest of the gyrus.

The superior longitudinal gyrus of the pig appears to be homologous to the middle and superior longitudinal gyri of the sheep and goat, and to the precentral gyrus of primates. The representation of the pig's musculature is not at all equally distributed. The

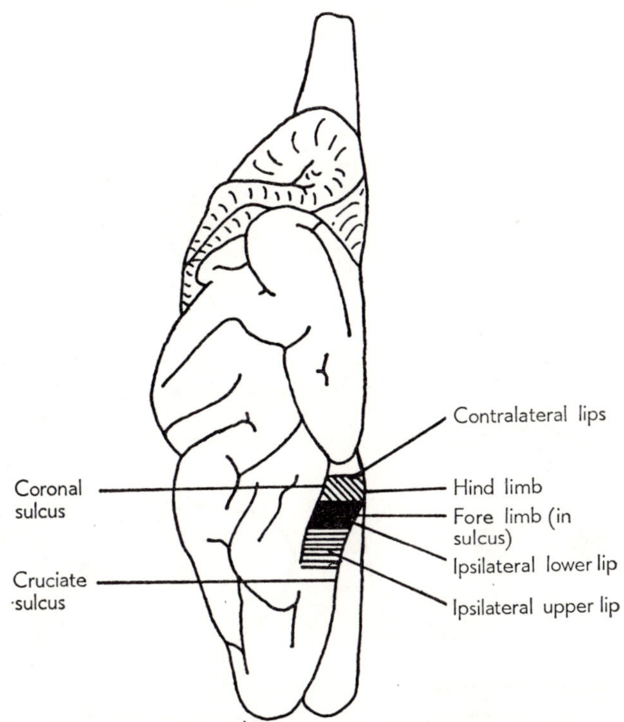

FIG. 11.1. Dorsal view of the cerebral cortex of the pig, illustrating the topography of the motor cortex and its organization relative to the body (from Breazile, Swafford & Thompson, 1966, by permission of *American Journal of Veterinary Research*).

greatest representation is of muscles of the lips, with distinct areas related to both ipsilateral and contralateral sides, and to upper and lower lips.

Sensory cortex

Adrian (1947) points out that an animal like the cat, which hunts its food and uses its forelimbs for capturing its prey, has a fairly

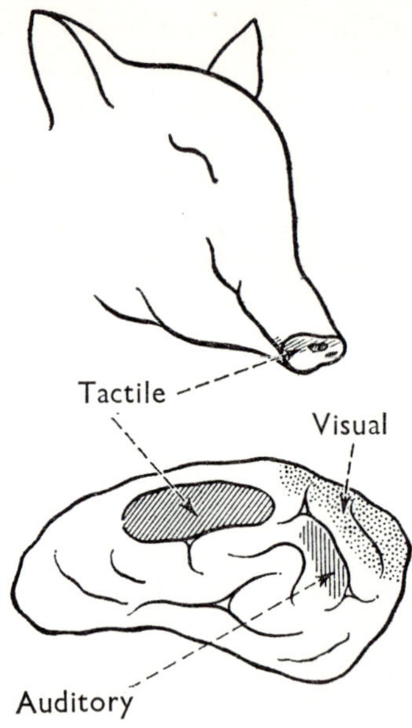

Fig. 11.2. Receiving areas in the cerebrum of the pig (from Adrian 1943*a*, by permission of *Lancet*).

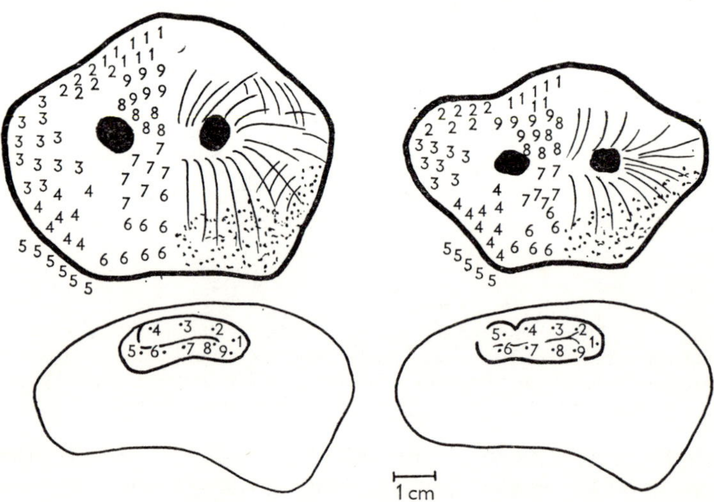

Fig. 11.3. Connections between snout receptors and cortical receiving area in two pigs. With the electrode at the point marked 2 on the brain (lower diagram), touch stimuli at all points marked 2 on the snout (upper diagram) gave discharges. Points marked 5 are on the lateral surface of the snout and margin of the upper lip (from Adrian, 1943*b*, by permission of *Brain*).

large receiving area in the brain cortex for messages from the forelimbs, whereas in the pig the tactile receiving area in the cortex appears to represent only the snout (Fig. 11.2), with the discharges coming from the contralateral half of the snout (Fig. 11.3). Adrian (1943*b*) estimated the ratio of the sensory surface to cortical receiving area, and concluded that the tactile discrimination of the pig's snout should be at least as great as that of the human hand.

Woolsey & Fairman (1946) later found face and fore-limb representation, both contralateral, in one cortical area, and face, fore-limb, and hind-limb, representation in another area. They found ipsilateral face areas as well as contralateral; and the largest potentials were produced by stimulation of the snout, limb apices, and the tail.

BEHAVIOUR: SENSORY CAPACITIES

Smell is the most important sense in the pig as determined from learning experiments. As regards the other special senses, there is some evidence that hearing is in general more important than vision (Hafez, Sumption & Jakway, 1962). Experiments on taste have demonstrated considerable individual variation in the behavioural reaction of pigs to saccharin solutions; the pig does, however, exhibit a preference for sugars described as sweet by man (Kare, Pond & Campbell, 1965).

The pig's snout

The pig's snout is a remarkable organ, not only in that it is intimately related to the animal's sense of smell, which is acute, and therefore important in the search for food and in awareness of other animals, but also because it is used as the animal's chief effector organ. The snout is to the pig what hands are to man. Adrian (1947) remarks: 'The pig's snout is its chief executive as well as its chief tactile organ, spade as well as hand, whereas the legs are little more than props for the body.'

The snout contains a cartilaginous core, which gives it mobility. Whereas the skin over the general body surface of the pig contains hair follicles and apocrine glands, the latter relatively few in number and apparently without thermoregulatory function, the skin over the snout is glabrous and contains eccrine sweat glands. It is astonishing what can be achieved by such an apparently sensitive 'chief executive organ'. Crandall (1964) writes of two wild boar

H

put into an animal house with an outside concrete run in the New York Zoo. Beginning from a small crack in the concrete surface of the run, within a few weeks the two pigs had by rooting with their snouts excavated and reduced to rubble concrete three to four inches thick.

Vision

The pig's eye is highly developed. The retina is rich in short, thick cones, and the retina, pupil, and lens are more like those of man than is the case for any of the other common farm animals. Klopfer (1966) has examined the characteristics of vision in pigs through the medium of visual learning experiments. He found that 8-month-old animals which had been fed from the same place every day did not respond to visual stimuli for food reward; however, animals given food scattered on the floor from the time of weaning could learn to discriminate between levels of brightness. Klopfer found that pigs respond to light within and perhaps beyond the range of wavelengths defined as light for the human, from below 420 up to 760 mμ. Learning experiments showed that pigs can distinguish between wavelengths (differences as small as 20 mμ) independently of brightness, which suggests that they can distinguish colours. The point of maximum sensitivity on the photopic (cone) visibility curve was at 575 mμ, instead of 540–50 which is the more common value for animals with colour vision. The scotopic (rod) visibility curve is similar to that for other animals (Fig. 11.4).

Klopfer concludes that pigs are functionally capable of a high level of visual performance, but that they do not use this ability, at least in feeding behaviour, if they are raised on stereotyped trough feeding. On the whole, their visual learning is similar to that of other higher mammals.

The effects of photoperiod on reproduction in some mammals are well known (Amoroso & Matthews, 1955; Fraps, 1962). Kločkova (1961) investigated the effects of photoperiod variation on sexual functions in the sow, by exposing animals to 18, 12, and 6 hours of light daily. The oestrous cycle was longest, and ovarian weight and follicle number lowest, in the 6 hr group. In growing pigs weight gains were highest in the 12 hr group, and bone growth least in the 6 hr group (Kločkova & Emme, 1964). Braude, Mitchell, Finn-Kelcey & Owen (1958), however, found no effect of light on weight gains in experiments in which the growth of pigs from 9 weeks of age onwards was measured under darkness throughout the 24 hours, or under continuous light, or with 10 or 14 hours of light per day. The different light treatments did not influence growth rate, efficiency of food utilization for growth, or the pattern of food consumption.

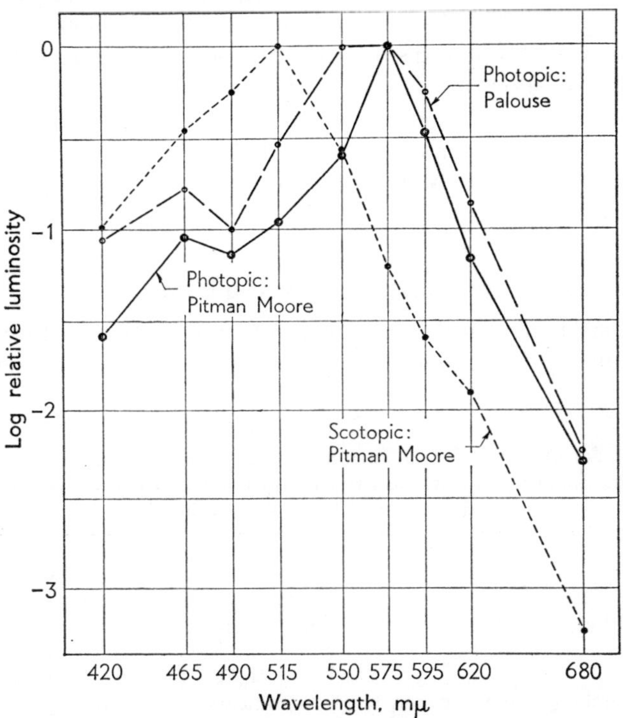

FIG. 11.4. Mean photopic visibility curves for seven Palouse and four Pitman-Moore pigs, and mean scotopic visibility for two Pitman-Moore pigs. The ordinate is expressed in units of relative brightness for each curve (from Klopfer, 1966, by permission of Frayn Printing Company, Seattle).

Hearing

Pigs have acute hearing, as any pig-keeper will testify. They can discriminate between speech signals, and quickly develop conditioned reflexes to words and short phrases (Konyukhova, 1966). In spite of this they are not perturbed by loud sounds. Bond, Winchester, Campbell & Webb (1963) sought an answer to the question whether loud aircraft sound subjected pigs to a level of stress which interfered with animal production. Normal farm pigs were used in the investigation, in conventional housing, and they were exposed to sequences of sound from flying aircraft and jet engines, at intensities up to 135db. The animals were almost invariably alarmed at the first exposure, but quickly became accustomed to loud sounds. Heart rate (recorded by radio telemetry) was increased significantly during the sound, but fell rapidly when the sound stopped. There was no indication that rate of growth, food intake, or food conversion were affected by exposure; mating, conception rate, and parturition were also not influenced. No injuries were caused to the pigs' ears, and there was no effect on the gross or microscopic structure of the adrenal and thyroid glands.

BEHAVIOUR: SOW AND LITTER

Sexual behaviour, and farrowing and nesting patterns, are described by Hafez *et al.* (1962). During lactation, the piglets tend

TABLE 11.1

The intake of milk by pigs sucking the anterior teats of sows compared with the mean for the whole litter (from Braude, 1954, by permission of Butterworth's Scientific Publications)

		Sow No. 1	*Sow No. 2*	*Sow No. 3*
Milk intake per pig per lactation, kg	mean	36·5	31·7	29·5
	left fore	40·4	42·9	25·6
	right fore	45·7	42·5	36·0
Milk intake per pig per day, g	mean	676	587	546
	left fore	748	794	474
	right fore	846	787	666
Milk intake at each suckling, g	mean	28·2	24·4	22·8
	left fore	31·2	33·1	19·8
	right fore	35·3	32·8	27·8
Milk consumed per g live weight gain in 21 days, g	mean	3·96	3·93	4·12
	left fore	3·88	3·98	4·27
	right fore	4·02	3·77	4·58

to feed at about one-hour intervals, although there are individual variations in frequency. Some of the measurable characteristics are referred to in Fig. 11.5.

Piglets develop a preference for particular teats during the first

hours after birth, and tend to return to the same teat at each feed.
When the sow rolls over, however, the teat order is upset, and
re-established only after competition among the piglets (McBride,
1963). The establishment of a teat order may have a stabilizing
influence in distributing food throughout the litter, and it may also
limit growth rate in individual new-born pigs. Any disturbance
may initiate suckling, particularly activity and noise on the part of
the litter, or sudden unusual sounds.

McBride, James & Wyeth (1965) found that social rank and
growth rate interacted continuously during the development of

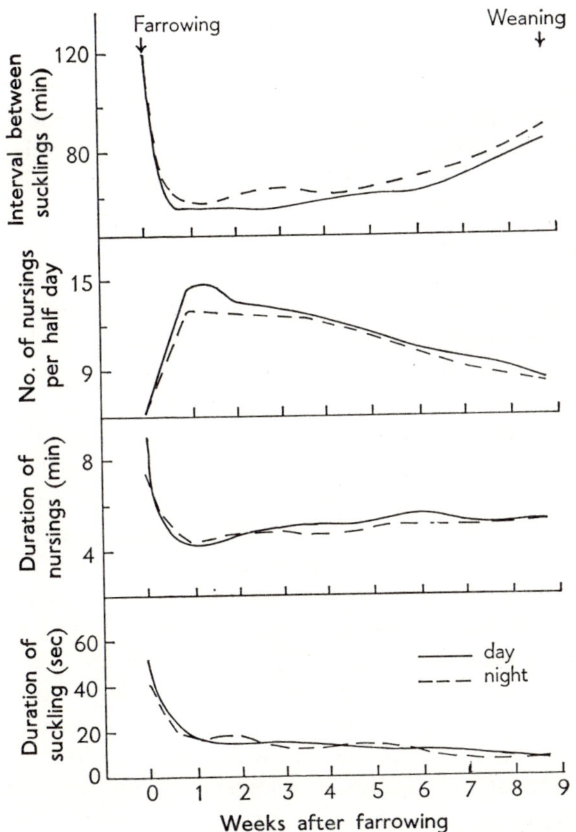

FIG. 11.5. The nursing behaviour of Middle White sows (5 litters)
during an entire lactation (based on results of Niwa, Yokoyama &
Otsuka, *Bull. nat. Inst. agric. Sci.* (Chiba), Ser. G., **1**, 135–50;
reproduced from Hafez, Sumption & Jakway, 1962, by permission
of Baillière, Tindall & Cassell).

young pigs. They found that both the teat order and the social order influenced the pig's growth to a considerable degree, with the social order being the more important. On the whole the more anterior teats yield the largest milk supply, but there is considerable variation between sows (Table 11.1).

BEHAVIOUR: THERMAL ENVIRONMENT

Pigs show a number of behavioural adaptations under temperature conditions which deviate from what is apparently a comfort zone for the animal. Huddling in the cold has already been discussed (Chapter 5); this type of behaviour has a considerable sparing effect on the energy required for maintenance in a cold environment. In addition to the obvious huddling of a number of pigs in a group, the single pig can modify heat loss to the environment quite effectively by changing posture. An example of this is the reduction of radiant heat loss from the new-born pig at 20°C as a result of a reduction in effective radiating area from 76% of the total surface area at 30°C to 67% at 20°C (Chapter 7). The animal's orientation to the wind probably influences convective heat loss, and the type of resting posture, whether relaxed or tense, influences conductive heat loss to the floor by affecting the area of contact (Chapter 8). In the Babraham pen calorimeter, the proximity of the pigs, one to another, clearly had a marked influence on the heat loss from the group (Chapter 5). Thus although heat losses at 20 and 30°C might have been similar, at 20°C the pigs rested in a huddle, whereas at 30°C they were spread out.

The pattern of defaecation and urination in the pigs in the calorimeter showed a definite relation to temperature. At 9 and 20°C, defaecation was confined to the 'dunging passage' area of the pen, leaving the sleeping and feeding area clean. At 30°C, however, excretion took place indiscriminately throughout the pen. This is in accordance with other experience that pigs kept on concrete floors under hot conditions tend to roll in voided urine, or any other water which is available. From Ingram's (1965*a*) measurements of rate of evaporation from wetted skin, this would be expected to provide considerable evaporative cooling (Chapter 6).

Preferred environments

The behavioural reactions mentioned so far have taken place within a limited type of environment which, apart from huddling, allowed very little scope either for the pig to choose one situation rather than another, or for it to operate on its environment and, in so doing, to change it. Heitman, Hahn, Bond & Kelly (1962) extended the range of variation open to the pig by offering a combination of shade, wallows, increased air movement, and an air-conditioned house, to pigs maintained in concrete pens under Californian summer conditions with an average mean temperature of 24°C, and with an average range of 15 to 35°C. They found that when shade was the only relief provided, it was used for about 80% of the time during the day-time; when wallows were provided as well, the time spent in the shade was reduced. The use of all the relief measures increased rapidly when the temperature rose above 21°C.

In such a situation, where choice is provided, pigs can discriminate readily between those conditions which impose a burden either on heat production or on heat loss, and those which do not. This applies also to baby pigs.

Thermocline

A simple thermocline was made in the form of a row of interconnecting hardboard compartments, each approximately 0·6 m wide, 0·65 m deep, and 0·65 m high (Mount, 1963c). The compartments communicated with each other through openings 25 cm high in the bottom of the dividing walls. The floors were 1 cm wire mesh, supported off the laboratory floor. Fans were arranged to blow air into the compartments, but air movement in the vicinity of the pig did not exceed 15 cm/sec. The air entering each compartment was heated to varying degrees, in this way producing a temperature range, in steps, from 23 to 37°C. The allocation of temperatures to the compartments could be varied.

Observations were made on young pigs between birth and 41 days of age, with a range of body weight from 0·9 to 9·7 kg. The animals were taken from the sow, and introduced into one of the compartments by a door in the front. The choice of the point of entry was made by lot, and the animals were introduced either one at a time or in groups of five pigs. The animals' positions were noted through small inspection holes at 15-minute intervals for 45 minutes in the case of single pigs, and for 60 minutes in the case of groups. With the single pigs, the animals had settled in the compartment of final choice within 30 minutes on 89% of occasions; with the groups, final choices had been made within 45 minutes on 72% of occasions. All five pigs of the groups were found together on 59% of occasions, and in a further 23% only one pig was separated from the group.

TABLE 11.2

Ambient temperatures preferred by young pigs, both singly and in groups of five; numbers choosing different temperatures (from Mount, 1963*c*, by permission of *Nature*)

Age of pigs *days*	*Ambient temperature °C*				*Mean ± S.E.* *temp. °C*
	23–28	*29–31*	*32–34*	*35–37*	
Single pigs					
< 1	1	2	5	2	32·3 ±0·94
1–7	8	16	4	0	29·3 ±0·44
8–41	16	33	11	6	29·8 ±0·64
Groups					
2–7	2	3	3	0	30·1 ±1·03
8–41	3	9	4	1	30·4 ±0·66

The choice of temperatures made by the pigs is given in Table 11.2. The preferred temperature is close to 30°C under these conditions, with a rather higher temperature for pigs less than one day old. Apart from this one difference, however, there is little effect of age. The critical temperature, from metabolic and skin temperature studies (Chapter 4), is about 34°C in the new-born, falling to 30°C and less at several weeks of age. On only 4 out of 28 occasions, however, did pigs less than one week old choose the 32–34°C compartment, and on 8 occasions they chose the 23–28°C compartment. The environment in the thermocline was such that air and wall temperatures were within 1°C, so that the operative temperature was close to the air temperature. The suggestion is that in order to be comfortable animals do not always have to be in the region of thermal neutrality as determined under standard conditions.

Operant conditioning

In the case of baby pigs in the thermocline, the animals were free to choose between a number of pre-existing situations. Another approach to the determination of preferred conditions is to allow the animal to perform a simple operation, such as pressing a lever, which results in a change in the environment. The frequency of pressing the lever, coupled with the kind of effect which the act produces, then gives information on the direction in which the animal would prefer its given environment to be changed and by how much.

Skinner (1938) developed experimental systems in which hungry or thirsty animals learned to press a lever in order to obtain food or water; with some types of drinking bowl used in normal farm practice, pigs learn this operation very rapidly. The techniques were extended so that rats kept in the cold learned to press a lever in order to obtain heat (Weiss, 1957a, b; Carlton & Marks, 1958; Weiss & Laties, 1961). Baldwin & Ingram (1967a) have used similar methods to study the pig's preferences. The pigs were trained by being placed in an experimental chamber at $-5°C$ for at least three sessions before measurements were begun (Fig. 11.6). In this cold environment the animals learned to make

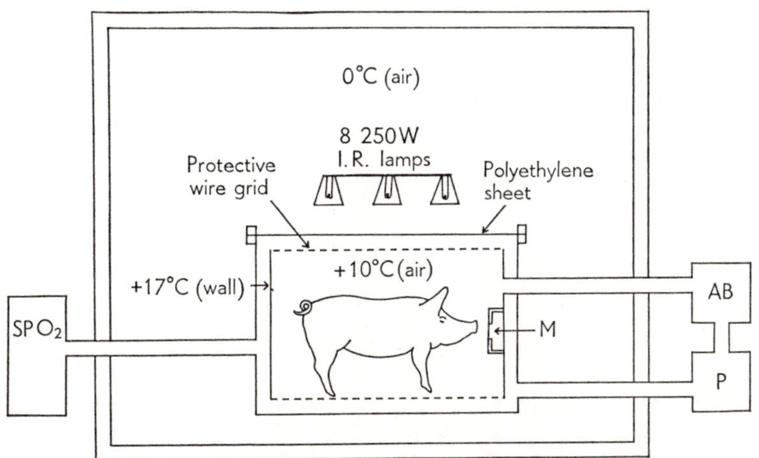

FIG. 11.6. Respiration chamber in a temperature-controlled room. By operating the heaters through the switch M, the pig has raised the air temperature of the chamber from its original 0°C to 10°C and raised the wall temperature to 17°C. SPO₂: spirometer filled with oxygen; P: pump; AB: carbon dioxide absorber (from Baldwin & Ingram, 1967a, by permission of *Physiology and Behaviour*).

a simple response (pressing a switch) which provided a short burst of infra-red heat as reinforcement. Individual pigs varied between 30 min and 10 hr in the time taken to learn to press the switch; at some point during this period the response rate would suddenly increase, and the animal would begin to make purposeful responses. As the environmental temperature was raised from

$-10°C$ to $+40°C$, the rate of response declined, markedly so above 20 to 30°C (Fig. 11.7).

The oxygen consumption rates of the animals were measured simultaneously with the measurement of their behavioural responses. It was found that the animals exposed to cold, but with access to heating operated by the switch, were able to keep their metabolic rates within the range found for the same animals

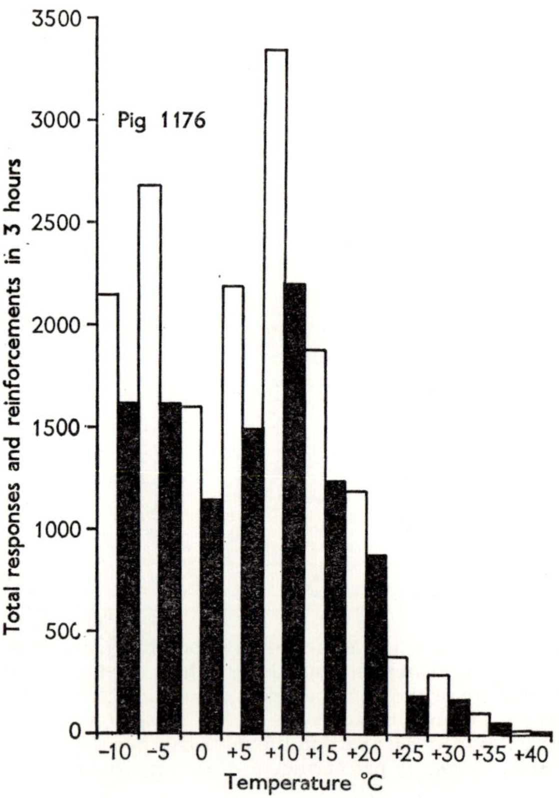

FIG. 11.7. The total responses and reinforcements (rewards in the form of bursts of infra-red heat) recorded for one pig during three-hour experimental sessions at environmental temperatures from -10 to $+40°C$. Responses, white; reinforcements, black. Responses made during a period of reinforcement had no effect, so that the number of responses exceeds the number of reinforcements (from Baldwin & Ingram, 1967a, by permission of *Physiology and Behaviour*).

exposed to a thermoneutral environment (Fig. 11.8). This occurred as the result of increased air and wall temperatures in the chamber, and was probably also due to the direct absorption by the animal of radiant heat; the mean operative temperature, with heaters being switched on and off, would be difficult to determine. The pigs also learned to switch off a draught of air in a cold environment. Under warm conditions this type of response decreased.

Pigs which were maintained on a relatively high food intake showed less inclination to obtain additional heat than animals kept on a low intake (Fig. 11.9). When pigs had been living at markedly

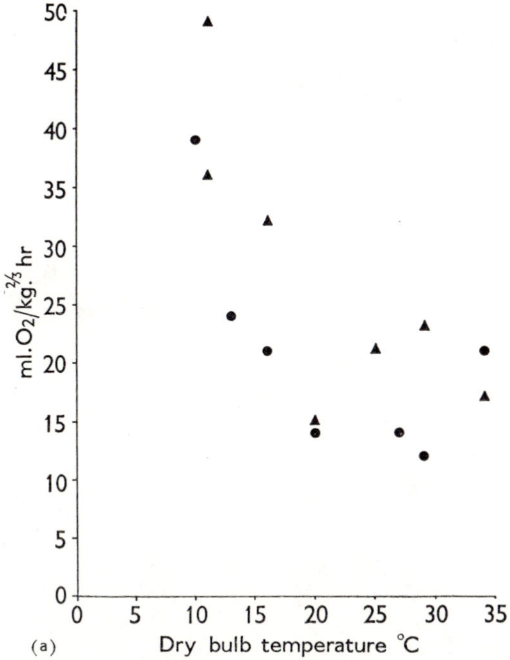

FIG. 11.8. Oxygen consumption of two pigs, ▲ and ●:
(*a*) in a chamber without radiant heat;
(*b*) in a chamber where the pigs could turn on radiant heaters by pressing a switch. The temperatures written beside the experimental points indicate the air temperature of the chamber at the time the pig was placed in it and before it operated the heaters. The abscissal temperatures were those at which oxygen consumption was measured, and which resulted from the pigs operating the heaters (from Baldwin & Ingram, 1967*a*, by permission of *Physiology and Behaviour*).

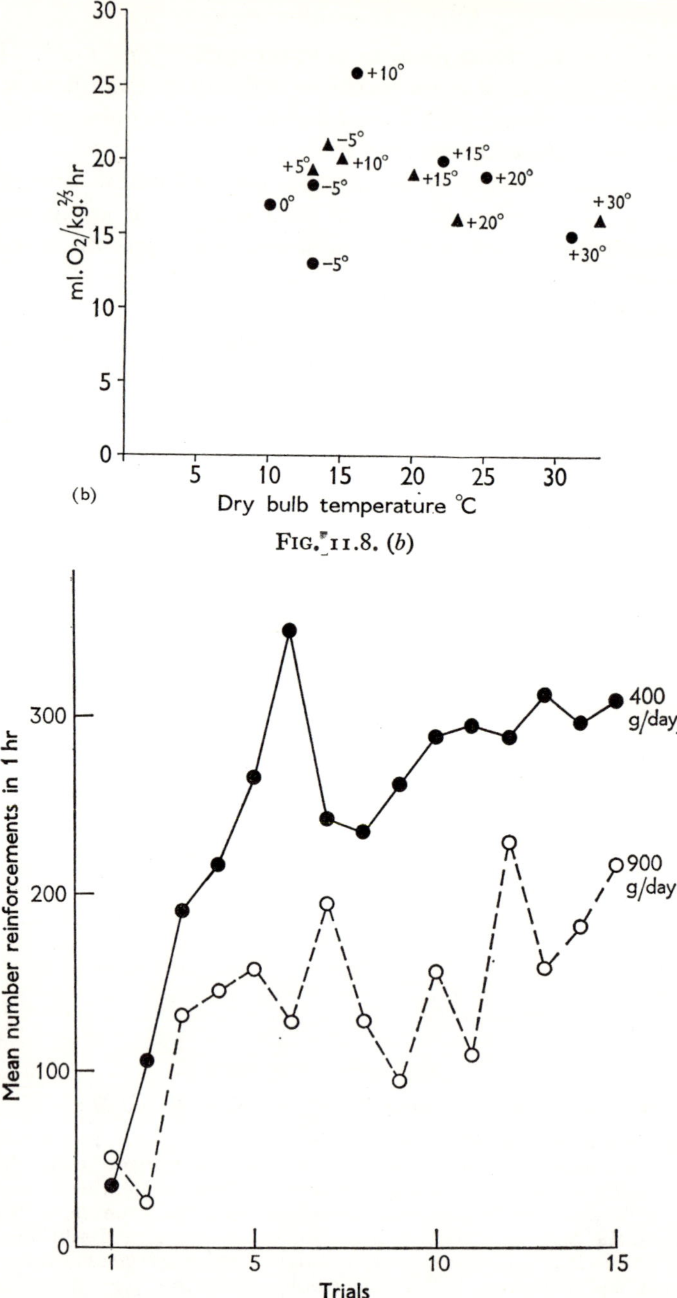

Fig. 11.8. (b)

Fig. 11.9. The effect of level of food intake on the number of heat reinforcements produced by pigs operating a switch (from Baldwin & Ingram, 1968a, by permission of *Physiology and Behaviour*).

different environmental temperatures (5 and 30°C), and were then exposed at 10°C in a situation where a bar could be pressed for additional radiant heat, the animals which had been living in the cold showed a consistently higher response rate (Fig. 11.10).

Using the technique which they had developed, Baldwin & Ingram (1967*b*) determined the response rates in pigs in which the hypothalamus was cooled by circulating cold liquid through an implanted thermode. They found that when the pre-optic region

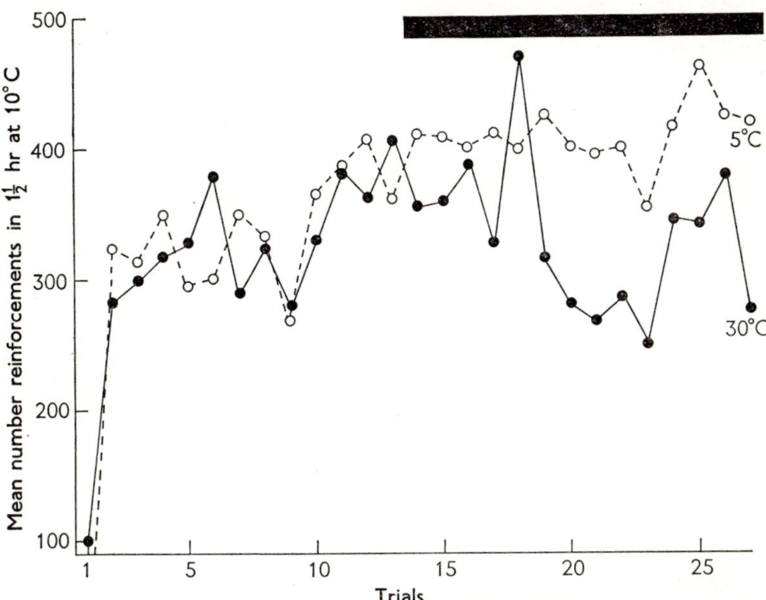

FIG. 11.10. Pigs lived at two temperatures, 5 and 35°C, during the period of three weeks indicated by the bar (trials were made on five days in each week). Trials were made at 10°C to determine the number of heat reinforcements obtained by the pigs pressing a switch (from Baldwin & Ingram, 1968*a*, by permission of *Physiology and Behaviour*).

of the hypothalamus was cooled, the rate at which the heaters were turned on increased, especially at temperatures in the region of the thermoneutral zone (Table 11.3). When the pre-optic region was warmed, the response rate decreased, but the effect was not as obvious as the increase seen during cooling. Another effect of cooling the hypothalamus is to decrease the high respiratory rate produced by a warm environment (Fig. 11.11).

In these ways, the use of the method of operant conditioning can contribute to understanding of the physiological mechanisms of thermoregulation and adaptation, in addition to providing information on the animal's behavioural responses and drives.

TABLE 11.3

The number of times heaters were turned on by a two-month-old pig, exposed to different environmental temperatures, during 30-minute periods of thermode cooling of the hypothalamus, and during similar periods preceding and following the cooling (from Baldwin & Ingram, 1967*b*, by permission of *Journal of Physiology*)

Environmental temperature, °C	Before cooling	During cooling	After cooling
0	64	122	68
	111	151	111
15	35	101	11
	148	262	140
20	52	84	2
	68	176	67
25	2	158	4
	17	52	0
30	0	48	0
	0	29	—
35	0	0	0
	9	0	0
Mean of all temperatures	42	99	35

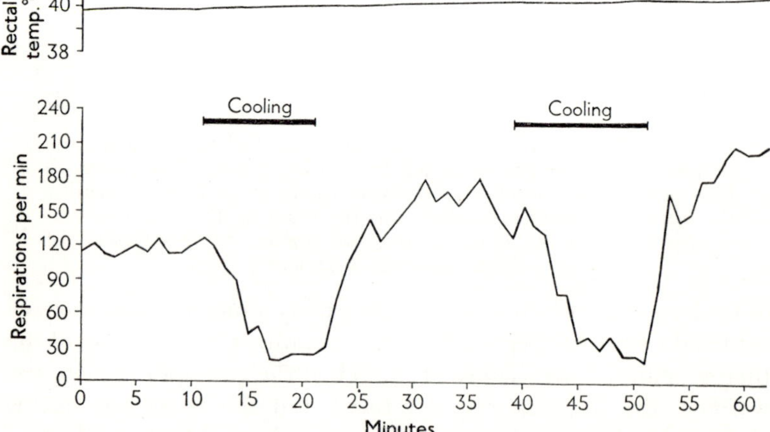

FIG. 11.11. The effect on respiratory rate in the pig of cooling the pre-optic region of the hypothalamus when the animal was panting at an environmental temperature of 40°C (from Baldwin & Ingram, 1968*b*, by permission of *Journal of Physiology*).

12

CLIMATIC PHYSIOLOGY IN RELATION TO PIG HUSBANDRY

MODERN livestock farming is concerned with providing large quantities of meat at as low a cost as possible. The economics of the situation has led to an increasing number of animals being housed under conditions of 'intensive husbandry', so that food intake, exercise, and environmental factors can be controlled. Pigs and poultry have been the two classes of livestock to be reared in very large numbers in this way, but the general method of controlled environment, as opposed to free-range existence, is also extended to sheep and cattle.

In the surroundings provided by intensive husbandry, the animals' freedom of choice of living conditions is obviously restricted, and consequently it is important that there should be a high degree of understanding of the effects of environmental factors on the animals' health, welfare, and productivity. The study of climatic physiology can contribute to this understanding (Mount, 1962*b*).

EXPERIMENTAL WORK AND FARM PRACTICE

The general requirement for farm practice in connection with pigs is to determine the range of conditions of environment and nutrition which allow maximum efficiency in the use of food for the growth of pigs grouped together in pens. Such a range of conditions would constitute a 'productivity plateau', in which productivity is maximal.

The productivity plateau in animal production is analogous to the thermoneutral zone which can be determined by physiological experiment. Over the extent of the plateau, productivity is maximal, and falls off on either side; in the thermoneutral zone, metabolic rate is minimum, and rises on either side. In both cases the rate of change on the colder side of the plateau or zone may be

229

expected to be higher in young pigs than in older pigs, and higher for single pigs than for groups of pigs. It is not to be expected that the plateau and the zone would *necessarily* coincide when both are assessed under equivalent conditions, and the relation of the plateau to the zone may vary from species to species. For example, man's efficient existence depends on the conditions in the micro-climate established in his clothing being close to those of thermal neutrality, but man can sweat more freely than any other animal and therefore he can tolerate incursions into much hotter con-ditions. It may be that the lower end of the productivity plateau in groups of pigs is below the critical temperature in so far as this applies to a group. Indeed, if the animals were housed at thermal neutrality, muscular activity, or feeding, could induce hyper-thermia because evaporative heat loss is inadequate for thermo-regulation in pigs under hot conditions (see Chapter 6). The provision of a temperature margin for activity may therefore be highly desirable.

TABLE 12.1

Factors which affect the use made by pigs of food, both for growth and maintenance

	Independent variables	*Dependent variables*
Nutrition:	plane	food conversion
	feeding frequency	growth rate
	protein content	body composition
	other dietary variation	carcass characteristics
		metabolic rate
Environment:	temperature, air and radiant,	
	steady or varying	
	humidity	
	air movement	
	floor, material,	
	space,	
	bedding	
	light cycle	
Animal:	breed	
	boar, castrate or female	
	age	
	body weight	
	group size	

The factors which affect the use made of food in growth can be listed as independent and dependent variables (Table 12.1). The independent variables are under the three headings of nutrition, environment, and animal, and define the conditions under which values for the dependent variables can be obtained. The level of nutrition can influence the result considerably; this is important when comparisons are attempted between pigs fed *ad libitum*, which is, for example, customary practice in the United States, and pigs fed restricted controlled amounts, which is common in Britain.

Optimal environment

Fig. 12.1 shows the apparent range of environmental temperature, 12–23°C, over which pigs appear to grow equally well, but

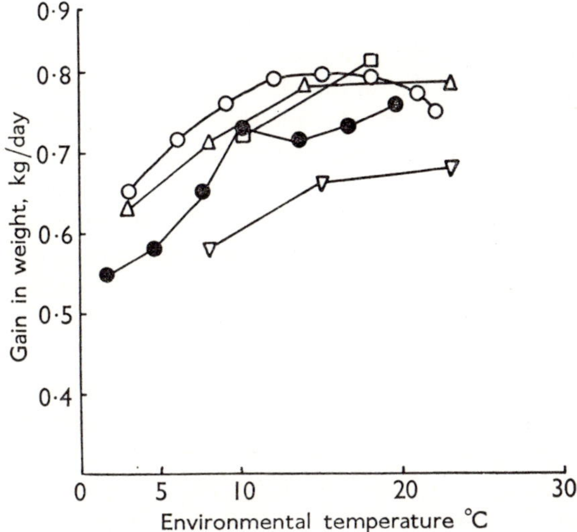

FIG. 12.1. Experiments on the gains in weight of pigs in relation to environmental temperature. The results are taken from:

□—□, Wöhlbier, Kirchgessner & Schneider (1958), *Arch. Tierernähr.* **8**, 32;
△—△ and ▽—▽, Sørensen (1960), Deutsche Akademie der Landwirtschaftswissenschaften, Berlin, *Tagungsberichte* Nr. 23, 97;
○—○, Comberg (1959), *Zuchtungskunde* **31**, 462;
●—●, Siegl (1960), *Archiv. f. Tierzucht* **3**, 188

(reproduced from Bianca & Blaxter, 1961, by permission of Eugen Ulmer, Stuttgart).

these results take no systematic account of the level of food intake. Sørensen (1962) suggested approximate limits of 12–19°C as bounding the zone of highest growth rate; Inglis & Robertson (1949) gave limits of 10–27°C based on a survey of pig housing.

These figures, however, refer only to air temperature and take no account of other components of the physical environment. Table 5.5 shows how bedding influences growth rate, presumably by influencing the micro-climate which directly impinges on the pig. McLagan & Thomson (1950) showed that where a dry insulated bed is provided it is possible to rear good pigs in a hut where the mean effective temperature is as low as 7°C (range 1 to 13°C) and in a piggery with a mean temperature of 3°C (−3 to 11°C). The type of flooring is of great importance; if this is well insulated, and particularly if bedding is provided, pigs can be reared in a very low mean environmental temperature, under conditions in which otherwise they would fail.

In general, however, the provision of a steady air temperature within the approximate limits of 15–20°C, with no draughts of air and with insulated floors, provide the conditions which call for the minimum maintenance energy on the part of groups of pigs in which each animal weighs 20 kg and more. This could be used as the definition of an optimal environment; to the farmer, however, the definition needs to be given in economic terms, so that what is optimal for growth and food utilization is not necessarily an economic optimum when the cost of providing such conditions is taken into account.

The large calorimeter described in Chapter 3 allows for a close approach to the ideal of pigs living in a farm-type situation but having measurements, of which they are unaware, made on them with the degree of precision expected of laboratory work. Such measurements provide the necessary information on growth and food utilization in relation to an environment controlled in a particular way. In the farm situation many more environmental transients, however, arise in the form of temperature fluctuations, varying draughts of air, and movements of personnel associated with the care of the animals.

'Comfort'

Irving (1964) has remarked that 'swine express their comfort or discomfort in terms that we understand'. Sainsbury (1954) assumed a similar understanding when, as the result of about 200 observations, he made suggestions regarding the comfort level of different ages of pig. The observations were made as subjective impressions of comfort, without reference to humidity or differences in husbandry, and indicate that as pigs grow older they are 'comfortable' at progressively diminishing temperatures, from about 21°C at birth to 5°C at 20 weeks of age and more; further, an increased air velocity has a greater effect on the younger pig.

In connection with these findings it is interesting that when Smith, C. V. (1964) made theoretical estimates (from results in the literature) of environmental demands for different ages of pig, he found that when Sainsbury suggested that the animals were showing signs of discomfort the demand is about 125% of the heat output in the comfort zone.

PROGENY TESTING AND ENVIRONMENT

In the pig industry there is continual striving to find the animal which will allow an increased efficiency of production. Greater efficiency can be achieved if growth can be accelerated, and if the amount of food required for a given live weight gain can be reduced.

It is necessary to carry out appropriate tests, based on objective standards and not on subjective judgement, if suitable animals are to be selected. Selection is most important in the case of boars, since one boar fathers many more piglets than one sow ever produces, a situation which is accentuated by the use of artificial insemination. This is the rationale of progeny testing, in which the growth and food utilization performance of a number of the boar's progeny are tested under controlled environmental conditions.

In order to carry out the test, the boar is mated to four sows, and two hogs and two gilts are chosen from each resulting litter. The pigs are put in pens at a controlled temperature; when they reach bacon weight the animals are slaughtered, and the carcasses are evaluated. Owing to possible environmental variation, useful comparisons between the progeny of different boars can be made only within one testing station, and preferably in the same season (Harrington, 1962).

The physical environment in the progeny testing stations is maintained with a minimum temperature and a controlled humidity. The effects of environment on the results of the tests are demonstrated when the pigs are housed in pens either singly, or two animals together; the difference between the results obtained by single and double penning is shown in Table 12.2. In double penning, rather less food is required per unit live weight gain, and the rate of growth is slightly higher. This is in accordance with the decreased heat losses experienced by pigs in a group when

TABLE 12.2

Single versus double penning in pig progeny testing. Means ±S.D. (by courtesy of Mr A. H. R. Pease and Pig Industry Development Authority)

	Single penned	*Double penned*
Number of pigs	206	212
Daily ration at 90 kg, in kg	3·19 ±0·230	3·20 ±0·226
Daily gain, kg	0·733 ±0·064	0·748 ±0·058*
Food conversion rate	3·052 ±0·251	3·004 ±0·163*
Days taken to reach 59 kg	52·80 ±3·98	51·44 ±5·20†
Food conversion rate up to 59 kg body weight	2·578 ±0·203	2·510 ±0·161‡

$* P<0.05.$ $† P<0.01.$ $‡ P\ 0.002.$

compared with single pigs (see Chapter 5). It was demonstrated by Prychodko (1958) that putting two mice together decreases the usual cold-induced rise in food consumption to a value below that found with single mice.

INVESTIGATIONS UNDER FARM CONDITIONS

Within the range of variation provided in the practical husbandry situation it is often difficult to establish criteria which will allow a useful level of discrimination between one set of conditions and another. This is partly because many factors become involved in tests made under practical conditions, with consequent interaction and loss of a clear effect in respect of any one variable.

Probably, however, the most important factor which leads to a reduction in differences of results between treatments lies in the adaptability of the pigs themselves. Such adaptability is apparent at both the physiological and behavioural levels. It is seen, for example, in the single pig in its posture in relation to wind, radiant heat loss, and a cold floor, and in groups of pigs in their huddling tendency in the cold. Measurements of heat loss from baby pigs to the floor has shown that the provision of a layer of straw on concrete is equivalent to raising floor temperature by 15°C (see

Chapter 8). It is therefore not altogether unexpected that in three houses markedly different from each other in respect of thermal insulation, but provided with straw bedding, the pigs should be similar in growth rate and food conversion (Cunningham & Rowell, 1963); the provision of straw extends the range of adaptation available to the animals.

Under more extreme conditions than those provided by the middle range of usual farm practice, however, and particularly with new-born pigs, the effects of environment do become evident. Boaz & Walters (1966) raised litters of new-born pigs to ten days of age at mean temperatures of 7 and 18°C. It was found that the piglets at 7°C were taking 15% more milk from the sow than those at 18°C; there were no significant differences between the body weights of the two groups at birth or at 10 days of age: the

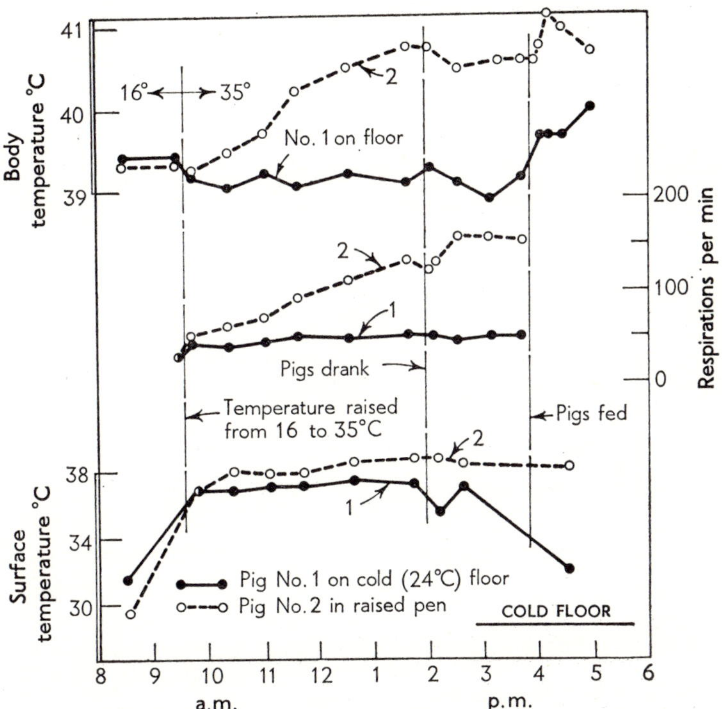

FIG. 12.2. Effect of a cool floor on body temperature, respiration, and surface temperature of a 45 kg pig in a 95°F (35°C) environment (from Bond, Heitman & Kelly, 1964, by permission of Commission Internationale du Génie Rural).

additional milk was taken to meet the increased maintenance demand of the colder environment. Larger pigs are more susceptible to high temperatures. Tonks & Smith (1965) compared the growth of pigs between about 20 and 90 kg body weight and kept at 28°C, 90% relative humidity, with their litter mates at 21°C, 70% relative humidity. Between about 20 and 70 kg body weight there was no significant effect associated with the environment, but between 70 and 90 kg body weight the pigs at the lower temperature showed a 10% higher growth rate. Heitman & Hughes (1949) and Heitman, Kelly & Bond (1958) have shown the effect of high temperatures in reducing food intake. They found temperatures in the region of 23°C to be optimal for pigs of 45 kg body weight, and 16°C for pigs of 160 kg, fed *ad libitum*. High temperatures lead to dirtying of pen floors; the difficulties associated with a high density of stock-

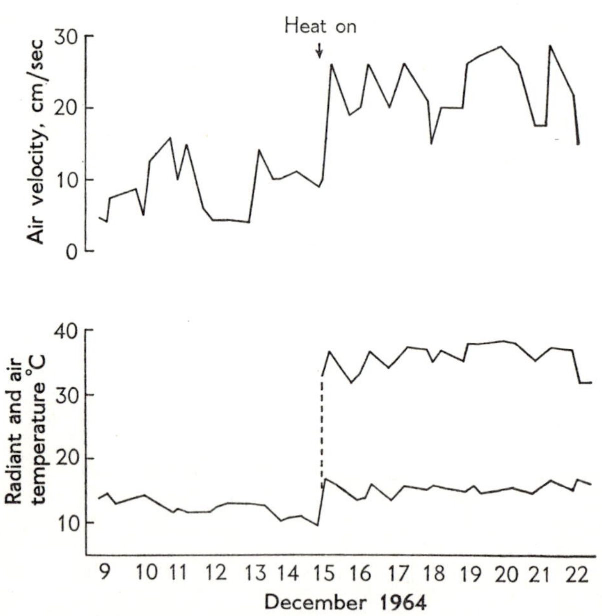

FIG. 12.3. Effect on air velocity (measured by katathermometer) in the vicinity of a heater lamp, suspended in a pig pen, when the lamp is switched on (at arrow). At this point, the dotted line indicates the rise in radiant temperature (measured by globe thermometer) to the value given by the upper temperature trace; the lower temperature trace gives air temperature (by courtesy of Mr F. R. Gilbert, and permission of the Pig Industry Development Authority).

ing animals in pens are accentuated under hot conditions (Gehlbach, Becker, Cox, Harmon & Jensen, 1966). Under hot conditions, pigs can lose a considerable amount of heat to a cool floor. Bond, Heitman & Kelly (1964) showed that a 45 kg pig in an air temperature of 35°C, but lying on a floor at 27°C, was able to maintain a body temperature about 1·5°C lower than a pig also exposed to 35°C but supported on a raised wire mesh floor. The surface temperature of the pig on the floor was also 1·5°C lower, and its respiration rate was reduced from 160 to 50 per minute (see Fig. 12.2).

Sometimes specific measures which are adopted in an attempt to ameliorate conditions serve only to make matters worse. A very common example of this is the frequent use of an infra-red heating lamp suspended over a litter of new-born pigs in an open creep area in a farrowing pen. Fig. 12.3 shows what happens to air movement, air temperature, and the globe thermometer temperature when such a lamp is switched on. Air movement is approximately doubled in the vicinity of the lamp, due to convection currents. Any beneficial effect of radiant heat in reducing the environmental thermal demand may be annulled by the increased draught, particularly if the air currents include colder air drawn in from outside the building. In this case the effect of a lowered air temperature is added to that of increased air speed.

FARM BUILDINGS

Considerable attention is paid by those concerned with animal husbandry to providing buildings which allow both the control of climate and the efficient management of livestock. The aim is to provide optimal conditions for production, in terms of growth and food utilization, and when this is impracticable on account of expense it is necessary to determine which compromise gives the greatest return for investment. The problem is complex, involving on the one hand requirements arising from the physiology, nutrition, reproduction, and behaviour of animals, and on the other the provision of physical facilities and systems of management so that production may be efficiently organized. Since so many different factors are interacting in the situation, it is to be expected that there are no simple universal solutions.

Buildings are designed to provide an internal climate which can be maintained within given limits independently of variations in the out-door climate. Because housing varies so much, there is considerable variation in the degree of protection from weather: the range extends from the open shed to the fully enclosed insulated house in which air temperature, air movement, humidity, and ventilation are controlled. The air temperature depends on the insulation, the heat produced by the animals, additional heat, ventilation rate, and the outside temperature. The increment of humidity in a house over the external level is due to evaporative loss from animals and from washing down, and depends on the ventilation rate (for practical aspects, see Sainsbury, 1963, 1967). Although the water content of the air in the house is higher than it is outside, the relative humidity may be lower since the air temperature is also higher. Any shelter reduces the mean air movement; air movement inside houses is a function of openings for natural ventilation, and depends also on the position and power of the fans used for ventilation. Air movement is often unevenly distributed, so that some animals may be in strong currents of cool air. Shelter also changes the radiant environment very considerably; in the place of a changing combination of cold sky, diffuse and direct solar radiation and, sometimes, radiation from hot ground, there is the relatively uniform internal temperature of the house. Only under the conditions of poorly insulated buildings in very hot or cold climates does the radiant environment inside a house have to be considered separately. In these cases extremes of radiant heat exchange can be reduced by artificial ceilings which tend to adopt air temperature.

Temperatures and humidities in pig houses have been investigated by Dick & Loader (1958). These authors related their observations to the insulation and ventilation of the houses, and their results allow calculations to be made for housing designs providing given internal temperatures and humidities. A knowledge of the animals' heat losses under various conditions of environment and nutrition makes such calculations applicable to a given situation (see Chapters 5 and 10).

In addition to the provision of a suitable crypto-climate (see Chapter 1), the design of pig houses has to allow for ease of handling and feeding animals and prevention of disease transmission. In growing pigs, much of the importance attached to the effects of environment lies in sparing food for growth by decreasing maintenance requirements. For the new-born pig, however, the interest lies more in ensuring the animal's survival. For this reason, the design of farrowing pens is influenced particularly by the need to achieve a low neonatal mortality. A common practice is to provide a 'creep', a railed-off warmed area into which baby pigs may move, and where the sow cannot follow. The creep provides for the piglets a micro-climate which is warmer than that required by the sow. It has been found under some conditions that the use of a 'farrowing crate', a built-up narrow space in which the sow is confined for parturition, gives a significantly lower mortality than the conventional open pen system (Robertson, Laird, Hall, Forsyth, Thomson & Walker-Love, 1966).

An extensive bibliography of farm buildings research is available (Harvey, 1959, 1962; Harvey & Easton, 1965, 1967).

DISEASE AND RESPONSES TO ENVIRONMENT

An animal's state of health can have effects on its reactions at both physiological and behavioural levels. An animal suffering from an infection does not gain weight at the same rate as another comparable but healthy animal fed on the same diet and level of food intake. The sick animal lacks the vigour characteristic of the healthy animal in its exploration of its environment; in the wild state, lack of well-being impairs the search for food, interferes with the reproductive drive, and militates against survival in a competitive world.

Pigs are subject to a number of diseases (Field, 1964; Anthony & Lewis, 1961; Dunne, 1964). In the case of the domestic pig, the sort of disease which may impair its performance depends on the regime under which it lives. Broadly, these regimes may be divided into two kinds: the open air and the indoor. Many of the difficulties which arise in intensive pig husbandry in houses with controlled environments do not occur in the open air situation; thus respiratory infections, which are so prevalent in housed pigs unless special measures are taken, are of little significance in pigs kept out-of-doors in a field, with huts for shelter. In the field, however, parasites can become a serious problem if pigs are kept for long periods continuously on one piece of ground.

In pig husbandry, all grades of housing exist ranging from the simplest type of shelter in a field, to a movable hut and open run in the field, or on concrete, then to pens in a house with access to the outside, and finally to a fully enclosed house in which the animals have no access either to the open air or to earth.

The open-air field situation

It was realized quite early in the development of modern intensive pig husbandry that one of the reasons why young pigs kept in huts with access to grass thrived so much more than animals kept in enclosed houses is that the indoor pigs suffered from anaemia. Sow's milk is relatively deficient in iron, and unless the pigs of a litter receive supplementary iron in some form, they suffer from an iron-deficiency anaemia. Out of doors enough additional iron is gained from rooting in the soil; indoors, the pigs must be given iron. The normal method is to inject a suitable iron preparation, or to give iron in some form by mouth.

From observations on piglets reared indoors, outdoors on pasture, outdoors on concrete or indoors with soil in the pen, however, it was found that those outdoors with soil grew faster than litters reared indoors or outdoors on concrete, although these last two groups were given iron pyrophosphate against anaemia (Barber, Braude & Mitchell, 1953). In winter, the difference between pigs outside on pasture and those kept indoors with access to soil was small. McLagan & Thomson (1950) had also demonstrated the superiority of the outdoor run (see Fig. 12.4). The results suggest that both vegetation and soil are beneficial to the piglet, rather than the temperature, sunlight, and fresh air associated with an outdoor environment.

The effect of exercise, sometimes coupled with an outdoor existence, has been investigated from time to time in respect of growth rate and body composition of pigs. Although there have been suggestions that exercise encourages growth rate and muscle formation (Gadzhiev, 1962; Ladrat, 1964), the results have so far not been convincing in economic terms.

Young sucking pigs may often be seen running about in the snow, behaviour which is apparently at variance with their marked susceptibility to chilling (Chapter 4). If the animals were to remain outside for long they would indeed tend to become hypothermic; in fact, following a period of brief exercise, they return to the warmth of the hut and a straw bed. If dry shelter is provided, exposure to wind, rain, and cold can be tolerated by the youngest pigs for brief periods.

An effect of climate which is more inescapable than cold is that of hot conditions to which larger pigs may be exposed. Mature sows are often kept outside in paddocks with small wooden arks for shelter. In a hot summer, such animals have no shade, and the temperature inside the arks rises to high levels. The animals will lie on the ground, breathing rapidly, and may suffer heat stroke if not provided with shade, wallows, or sprays. Such provision is normal in hot countries, and is essential in an animal with no effective sweating mechanism.

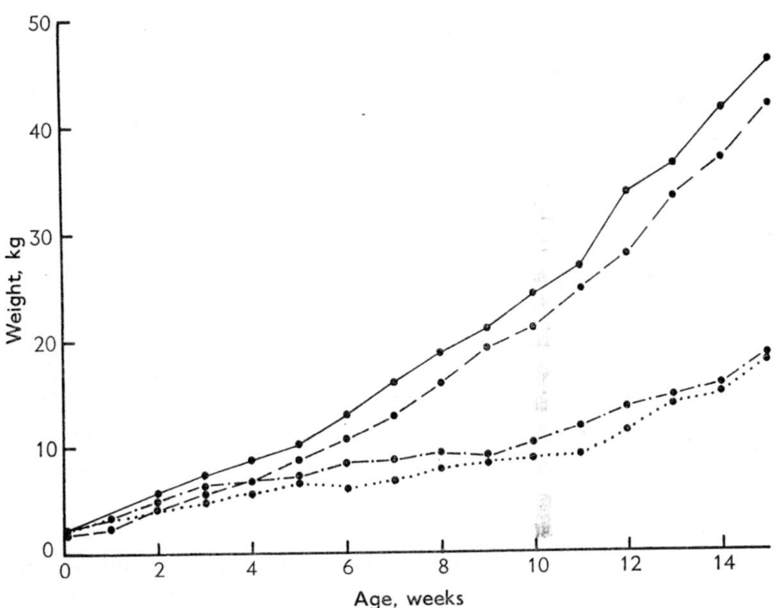

FIG. 12.4. Average weights of pigs in four types of housing from birth to 15 weeks.

——— pigs from hut with outdoor run;
—·—· pigs from hut with indoor run;
— — — pigs from pen with wooden floor;
. . . . . pigs from pen with uninsulated floor

(from McLagan & Thomson, 1950, by permission of *Journal of Agricultural Science*).

The enclosed house situation

Pigs living entirely in houses are subjected to varying degrees of control of their physical environment. Some control of air temperature may be effected, but this is sometimes accompanied by damp conditions and continued exposure to cold, uninsulated floors. The mortality in new-born pigs under such conditions is high, and may be associated with various disease syndromes including 'baby pig disease' (see Chapter 2) which is due to hypoglycaemia, and liver disease (Howie, 1951). In older pigs exposed to an ill-controlled environment there is failure to gain weight at the rate expected of healthy animals on food intakes of high level and quality, and respiratory infections and enteritis are prevalent.

A common situation which arises with successive groups of pigs raised in a building is the progressive decline in productivity, in the form of growth rate and food conversion, during the course of time. The evacuation of the building, followed by cleaning and sterilization of the interior, are sometimes followed by a return to the original level of productivity when the house is put back into use again. In the absence of specific findings, this phenomenon is attributed to an increase in bacterial populations and the accompanying 'sub-clinical' forms of disease.

If air temperature and humidity are both kept at high levels, there is an improvement in growth on a given food intake and the morbidity associated with respiratory infections declines. This type of husbandry has been investigated by Gordon (1963*a*, *b*), who found that, under high humidity conditions, water droplets in the air increased in size, and the bacterial count in the air fell.

Morrison, Heitman, Bond & Finn-Kelcey (1966) found no differences in the health or general condition of pigs depending on humidity. They considered that the beneficial effects of high humidity might occur only where respiratory disease existed. An important practical conclusion was that evaporative cooling of hot dry air at the expense of increasing the humidity can be recommended, since the benefit of lower air temperature would more than compensate for any adverse effect due to a higher humidity. Morrison, Bond & Heitman (1967) have recently derived a relation between the weight-gains of 60–70 kg pigs and the combination of humidity and any high temperature above 22°C. This relation allows the prediction of the effects of these two environmental factors on the rate of weight gain in pigs fed *ad libitum*.

When pigs are housed in enclosed buildings the air-space can become contaminated by various gases, including ammonia, hydrogen sulphide, and carbon dioxide, derived from the breakdown of excreta (Day, Hansen & Anderson, 1965). The concentrations of gases depend to some extent on the type of house employed, whether it has solid floor dunging passages, or slatted floors, which allow urine and faeces to fall through the floor to a pit or trench below, from which periodic collection is undertaken.

REQUIREMENTS FOR INTENSIVE HUSBANDRY UNDER CONTROLLED CONDITIONS

When pigs are kept in confined quarters under conditions which fall short of the environmental optimum, they tend to be more prone to disease than pigs kept in more extensive conditions out-of-doors. There is now an increasing tendency for pigs to be reared free of those common diseases which are the most troublesome. Some herds are maintained free of enzootic pneumonia, a troublesome and insidious disease which is apparently spread nearly always only by direct pig-to-pig contact (Whittlestone, 1960; Goodwin, 1963). Other herds, termed 'minimal disease herds', are raised from 'specific-pathogen-free' (SPF) pigs, which are obtained initially by hysterectomy near term, the piglets being passed to sterile containers from the uterus through antiseptic traps (Betts, Lamont & Littleworth, 1960).

The combination of freedom from infective agents, environmental conditions approaching the optimum for minimum maintenance energy requirements, and a nutritionally satisfactory diet, may be expected to lead to as high a utilization of food intake for growth as can be expected. It is under these conditions that the animals need highly skilled attention if they are to thrive. The removal of the animals' relative freedom by confinement under a given regime means that they may be unable to adapt satisfactorily to imposed environmental stresses, and the lack of adaptation plus the frustration of natural tendencies can lead to deviant behaviour patterns such as tail-biting. It is evident that the intensive husbandry of pigs calls for the exercise of a high level of animal care and an understanding of the animal's physiological and behavioural requirements.

REFERENCES

Adamsons, K., Jr., Gandy, G. M. & James, L. S. (1965). The influence of thermal factors upon oxygen consumption of the new-born human infant. *J. Pediat.* **66,** 495–508.

Adrian, E. D. (1943*a*). Sensory areas of the brain. *Lancet,* **ii,** 33–36.

Adrian, E. D. (1943*b*). Afferent areas in the brain of ungulates. *Brain,* **66,** 89–103.

Adrian, E. D. (1947). *The physical background of perception.* Oxford: Clarendon.

Agricultural Research Council (1967). *The nutrient requirements of farm livestock. No. 3 Pigs: technical reviews and summaries.* London: Her Majesty's Stationery Office.

Alexander, G. (1961*a*). Temperature regulation in the new-born lamb. II. A climatic respiration chamber for the study of thermoregulation. *Aust. J. agric. Res.* **12,** 1139–1151.

Alexander, G. (1961*b*). Temperature regulation in the new-born lamb. III. Effect of environmental temperature on metabolic rate, body temperatures, and respiratory quotient. *Aust. J. agric. Res.* **12,** 1152–1174.

Alexander, G. (1962). Temperature regulation in the new-born lamb. V. Summit metabolism. *Aust. J. agric. Res.* **13,** 100–121.

Alexander, G. & Brook, A. H. (1960). Loss of heat by evaporation in young lambs. *Nature, Lond.* **185,** 770–771.

Alexander, G. & McCance, I. (1958). Temperature regulation in the new-born lamb. *Aust. J. agric. Res.* **9,** 339–347.

Allen, A. D. & Lasley, J. F. (1954). Influence of season of birth on growth rate and survival of pigs. *J. anim. Sci.* **13,** 955.

Allen, A. D., Lasley, J. F. & Tribble, L. F. (1959). Milk production and related performance factors in sows. *Missouri Agric. Expt. Sta. Res. Bull.* 712.

Amoroso, E. C. & Matthews, L. H. (1955). The effect of external stimuli on the breeding-cycle of birds and mammals. *Br. med. Bull.* **11,** 87–92.

Anthony, D. J. & Lewis, E. F. (1961). *Diseases of the pig,* 5th edition. London: Ballière, Tindall and Cox.

Armsby, H. P. & Fries, J. A. (1903). The available energy of 'Timothy' Hay. *U.S. Dept. Agriculture. Bull.* 51.

Aschoff, J. & Wever, R. (1958). Kern und Schale im Wärmehaushalt des Menschen. *Naturwissenschaften* **20,** 477–485.

Atwater, W. O. & Benedict, F. G. (1905). *A respiration calorimeter with appliances for the direct measurement of oxygen.* Carnegie Inst. Publ. No. 42.

Auguet, A. & Lefèvre, J. (1929). Nouvelle chambre calorimetrique du laboratoire de bioenergitique. *C. r. hebd. Seanc. Acad. Sci.,* Paris. **100,** 251–253.

Bailey, C. B., Kitts, W. D. & Wood, W. J. (1956). The development of the digestive enzyme system of the pig during its pre-weaning phase of growth. *Can. J. agric. Sci.* **36,** 51–58.

Baldwin, B. A. & Ingram, D. L. (1967*a*). Behavioural thermoregulation in pigs. *Physiol. Behav.* **2,** 15–21.

Baldwin, B. A. & Ingram, D. L. (1967*b*). The effect of heating and cooling the hypothalamus on behavioural thermoregulation in the pig. *J. Physiol., Lond.* **191,** 375–392.

Baldwin, B. A. & Ingram, D. L. (1968*a*). The effects of food intake and acclimatization to temperature on behavioural thermoregulation in pigs and mice. *Physiol. Behav.* in the press.

Baldwin, B. A. & Ingram, D. L. (1968*b*). The influence of hypothalamic temperature and ambient temperature on thermoregulatory mechanisms in the pig. *J. Physiol., Lond.* in the press.

Barber, R. S., Braude, R. & Mitchell, K. G. (1953). Indoor v. outdoor rearing of piglets. *Proc. Nutr. Soc.* **12,** xiii.

Barber, R. S., Braude, R. & Mitchell, K. G. (1955). Studies on milk production of Large White pigs. *J. agric. Sci., Camb.* **46,** 97–118.

Barić, I. (1953). La consommation d'oxygène du rat nouveau-né au cours du jeûne. *Bull. de l'Acad. Serbe Sci.* **12,** 71–76.

Barnett, S. A. (1959). The skin and hair of mice living at a low environmental temperature. *Q. Jl exp. Physiol.* **44,** 35–42.

Barnett, S. A. & Mount, L. E. (1967). Resistance to cold in mammals. In *Thermobiology*, ed. A. H. Rose. London: Academic Press.

Bazett, H. C., Love, L., Newton, M., Eisenberg, L., Day, R. & Forster, II, R. (1948). Temperature changes in blood flowing in arteries and veins in man. *J. appl. Physiol.* **1,** 3–19.

Becker, D. E., Ullrey, D. E. & Terrill, S. W. (1954). A comparison of carbohydrates in a synthetic milk diet for the baby pig. *Archs Biochem. Biophys.* **48,** 178–183.

Becker, D. E., Ullrey, D. E., Terrill, S. W. & Notzold, R. A. (1954). Failure of the new-born pig to utilize dietary sucrose. *Science,* **120,** 345–346.

Bedford, T. (1946). *Environmental warmth and its measurement.* Medical Research Council, London: H.M.S.O.

Benedict, F. G. (1938). *Vital energetics.* Carnegie Inst. Publ. 503, Washington.

Benzinger, T. H., Huebscher, R. G., Minard, D. & Kitzinger, C. (1958). Human calorimetry by means of the gradient principle. *J. appl. Physiol.* **12,** S1–28.

Benzinger, T. H. & Kitzinger, C. (1949). Direct calorimetry by means of the gradient principle. *Rev. Sci. Instrum.* **20,** 849–860.

Benzinger, T. H. & Kitzinger, C. (1963). Gradient layer calorimetry and human calorimetry. In *Temperature, its measurement and control in science and industry*, Vol. III, Part 3, ed. J. D. Hardy, pp. 87–109. New York: Reinhold.

Berry, I. L. & Shanklin, M. D. (1961). Environmental Physiology and Shelter Engineering. LXIV. Physical factors affecting thermal insulation of livestock hair coats. *Res. Bull. 802, Mo. agric. Exp. Sta.*

Betts, A. O., Lamont, P. H. & Littlewort, M. C. G. (1960). The production by hysterectomy of pathogen-free, colostrum-deprived pigs and the foundation of a minimal-disease herd. *Vet. Rec.* **72,** 461–468.

Bianca, W. (1962). Relative importance of dry- and wet-bulb temperatures in causing heat stress in cattle. *Nature, Lond.* **195,** 251–252.

Bianca, W. & Blaxter, K. L. (1961). The influence of the environment on animal production and health under housing conditions. *VIII Internat. Congr. Animal Husb., Hamburg.* Stuttgart: Eugen Ulmer.

Blaxter, K. L. (1962*a*). *The energy metabolism in ruminants.* Charles C. Thomas, Springfield, Illinois.

Blaxter, K. L. (1962*b*). The fasting metabolism of adult wether sheep. *Br. J. Nutr.* **16,** 615–626.

Blaxter, K. L. (1964). The effect of outdoor climate in Scotland on sheep and cattle. *Vet. Rec.* **76,** 1445–1455.

Blaxter, K. L. (ed.). (1965). *Energy Metabolism.* London: Academic Press.

Blaxter, K. L., Graham, N. McC. & Wainman, F. W. (1959). Environmental temperature, energy metabolism and heat regulation in sheep. III. The metabolism and thermal exchanges of sheep with fleeces. *J. agric. Sci., Camb.* **52,** 41–49.

Blaxter, K. L., Graham, N. McC., Wainman, F. W. & Armstrong, D. G. (1959). Environmental temperature, energy metabolism and heat regulation in sheep. II. The partition of heat losses in closely clipped sheep. *J. agric. Sci., Camb.* **52,** 25–40.

Blaxter, K. L., Joyce, J. P. & Wainman, F. W. (1963). Effect of air velocity on the heat losses of sheep and cattle. *Nature, Lond.* **198,** 1115–1116.

Blaxter, K. L. & Wainman, F. W. (1961). Environmental temperature and the energy metabolism and heat emission of steers. *J. agric. Sci., Camb.* **56,** 81–90.

Blaxter, K. L. & Wainman, F. W. (1966). The fasting metabolism of cattle. *Br. J. Nutr.* **20,** 103–111.

Bligh, J. (1966). The thermosensitivity of the hypothalamus and thermoregulation in mammals. *Biol. Rev.* **41,** 317–367.

Blum, H. F. (1961). Does the melanin pigment of human skin have adaptive value? *Q. Rev. Biol.* **36,** 50–63.

Boaz, T. G. & Walters, A. A. (1966). Weight and growth rate. *Pig Farming* **14,** 45–47.

Bond, J., Winchester, C. F., Campbell, L. E. & Webb, J. C. (1963). Effects of loud sounds on the physiology and behaviour of swine. *U.S. Dept. Agric., Tech. Bull. 1280,* Washington, D.C.

Bond, T. E., Heitman, Jr., H. & Kelly, C. F. (1964). Physiological response time of thermally stressed swine to several cooling media. *VIth. Internat. Contr. Agr. Engrg, Lausanne,* Vol. 1, 356–370.

Bond, T. E., Heitman, Jr., H. & Kelly, C. F. (1965). Effects of increased air velocities on heat and moisture loss and growth of swine. *Trans. Am. Soc. agric. Engrs* **8,** 167–169 and 174.

Bond, T. E., Kelly, C. F. & Heitman, Jr., H. (1952). Heat and moisture loss from swine. *Agr. Engrg, St. Joseph, Mich.* **33,** 148–152.

Bond, T. E., Kelly, C. F. & Heitman, Jr., H. (1963). Effect of diurnal temperature on heat loss and well-being of swine. *Trans. Am. Soc. agric. Engrs* **6,** 132–135.

Bond, T. E., Kelly, C. F. & Heitman, Jr., H. (1967). Physiological response of swine to cycling environmental conditions. *Anim. Prod.* **9**, 453–462.

Bond, T. E., Kelly, C. F., Morrison, S. R. & Pereira, N. (1967). Solar, atmospheric and terrestrial radiation received by shaded and unshaded animals. *Trans. Am. Soc. agric. Engrs* **10**, 622–625, 627.

Bonsma, J. C. (1943). Influence of colour and coat cover on adaptability of cattle. *Farming in South Africa*. February no., 1943.

Bowland, J. P. (1966). Swine milk composition—a summary. In *Swine in biomedical research*, ed. L. K. Bustad & R. O. McClellan, p. 100. Battelle Memorial Institute.

Braude, R. (1954). Pig Nutrition. In *Progress in the Physiology of Farm Animals*, ed. J. Hammond, Vol. 1, p. 48. London: Butterworths.

Braude, R., Mitchell, K. G., Finn-Kelcey, P. & Owen, V. M. (1958). The effect of light on fattening pigs. *Proc. Nutr. Soc.* **17**, xxxviii.

Breazile, J. E., Swafford, B. C. & Thompson, W. D. (1966). Study of the motor cortex of the domestic pig. *Am. J. vet. Res.* **27**, 1369–1373.

Brockway, J. M., McDonald, J. D. & Pullar, J. D. (1965). Evaporative heat-loss mechanisms in sheep. *J. Physiol., Lond.* **179**, 554–568.

Brody, S. (1945). *Bioenergetics and growth*. New York: Reinhold.

Brody, S. & Kibler, H. H. (1944). Resting energy metabolism and pulmonary ventilation in growing swine. *Mo. Agr. Exp. Sta. Res. Bull.* 380.

Brooks, C. C., Fontenot, J. P., Vipperman, P. E., Thomas, H. R. & Graham, P. P. (1964). Chemical composition of the young pig carcass. *J. anim. Sci.* **23**, 1022–1026.

Brooks, C. C., Thomas, H. R., Kelley, R. F., Graham, P. P. & Allen, L. B. (1964). *Virginia Agric. Exp. Sta. Tech. Bull.* 176, pp. 15.

Burton, A. C. & Edholm, O. G. (1955). *Man in a cold environment*. London: Edward Arnold.

Bustad, L. K., Horstman, V. G. & England, D. C. (1966). Development of Hanford miniature swine. In *Swine in Biochemical Research*, pp. 769–774. eds. L. K. Bustad & R. O. McClellan. Battelle Memorial Institute.

Butchbaker, A. F. & Shanklin, M. D. (1964). Partitional heat losses of new-born pigs as affected by air temperature, absolute humidity, age, and body weight. *Trans. Am. Soc. agr. Engrs.* **7**, 380–383.

Cairnie, A. B. (1958). *The study of the heat output of the young pig under various conditions*. Ph.D. Thesis, University of Aberdeen.

Cairnie, A. B. & Pullar, J. D. (1959). An investigation into the efficient use of time in the calorimetric measurement of heat output. *Br. J. Nutr.* **13**, 431–439.

Cannon, P. & Keatinge, W. R. (1960). The metabolic rate and heat loss of fat and thin men in heat balance in cold and warm water. *J. Physiol., Lond.* **154**, 329–344.

Capstick, J. W. (1921). A calorimeter for use with large animals. *J. agric. Sci., Camb.* **11**, 408–431.

Capstick, J. W. & Wood, T. B. (1922). The effect of change of temperature on the basal metabolism of swine. *J. agric. Sci., Camb.* **12**, 257–268.

I

Carlton, P. L. & Marks, R. A. (1958). Cold exposure and heat reinforced operant behaviour. *Science* **128**, 1344.

Cherian, A. G. (1962). Metabolism as a function of age and weight in frog. *Acta physiol. pharmacol. Neerlandica* **11**, 443–456.

Crandall, L. S. (1964). *The management of wild mammals in captivity.* Univ. of Chicago Press.

Crawford, A. (1779). *Experiments and observations on animal heat and the inflammation of combustible bodies, being an attempt to resolve these phenomena into a general law of nature.* Murray and Sewell, London.

Crawford, A. (1788). *Experiments and observations on animal heat and the inflammation of combustible bodies.* 2nd edition. Johnson, London.

Cresswell, E. & Thomson, W. (1964). An introductory study of air flow and temperature at grazing-sheep heights. *Emp. J. exp. Agr.* **32**, 131–135.

Cross, K. W., Gustavson, Jean, Hill, June R. & Robinson, D. C. (1966). Thermoregulation in an anencephalic infant as inferred from its metabolic rate under hypothermic and normal conditions. *Clin. Sci.* **31**, 449–460.

Cunha, T. J. (1966). Nutritional requirements of the pig. In *Swine in Biomedical Research*, eds. L. K. Bustad & R. O. McClellan, pp. 681–696. Battelle Memorial Institute.

Cunningham, J. M. M. & Rowell, J. G. (1963). Is insulation over-rated? *Pig Farming.* March (1963), pp. 41 and 43.

Cuthbertson, A. & Pomeroy, R. W. (1962). Quantitative anatomical studies of the composition of the pig at 50, 68 and 92 kg carcass weight. II. Gross composition and skeletal composition. *J. agric. Sci., Camb.* **59**, 215–223.

Dahlqvist, A. (1961). Intestinal carbohydrases of a new-born pig. *Nature, Lond.* **190**, 31–32.

David, L. T. (1932). Histology of the skin of the Mexican hairless swine (Sus scrofa). *Am. J. Anat.* **50**, 283–292.

Davidson, H. R. (1954). *The production and marketing of pigs.* London: Longmans, Green.

Davis, T. R. A., Johnston, D. R., Bell, F. C. & Cremer, B. J. (1960). Regulation of shivering and non-shivering heat production during acclimatization of rats. *Am. J. Physiol.* **198**, 471–475.

Dawes, G. S. (1961). Oxygen consumption and hypoxia in the new-born animal. In *Somatic stability in the newly born*, Ciba Found. Symp. eds. G. E. W. Wolstenholme & M. O'Connor, pp. 170–191. London: Churchill.

Dawes, G. S., Jacobson, H. N., Mott, J. C. & Shelley, H. J. (1960). Some observations on foetal and new-born Rhesus monkeys. *J. Physiol., Lond.* **152**, 271–298.

Dawes, G. S. & Shelley, H. J. (1968). Physiological aspects of carbohydrate metabolism in the foetus and new born. In *Carbohydrate Metabolism and its Disorders*, eds. F. Dickens, P. J. Randle and W. J. Whelan, vol. II, pp. 87–121. London, New York: Academic Press.

Dawkins, M. J. R. & Hull, D. (1964). Brown adipose tissue and the response of new-born rabbits to cold. *J. Physiol., Lond.* **172**, 216–238.

Day, D. L., Hansen, E. L. & Anderson, S. (1965). Gases and odors in confinement swine buildings. *Trans Am. Soc. agric. Engrs.* **8**, 118–121.

Day, R. & Hardy, J. D. (1942). Respiratory metabolism in infancy and in childhood. 26. A calorimeter for measuring the heat loss of premature infants. *Am. J. dis. Child.* **63**, 1086–1095.

Deighton, T. (1924). The basal metabolism of a growing pig. *Proc. Roy. Soc.* **B.95**, 340–355.

Deighton, T. (1929). A study of the metabolism of two breeds of pig. *J. agric. Sci., Camb.* **19**, 140–184.

Deighton, T. (1932). The determination of the surface area of swine and other animals. *J. agric. Sci., Camb.* **22**, 418–449.

Deighton, T. (1935). A study of the fasting metabolism of various breeds of pig. II. Body temperature measurements. *J. agric. Sci. Camb.* **25**, 180–191.

Deighton, T. (1937). A study of the fasting metabolism of various breeds of hog. III. Metabolism and surface area measurements. *J. agric. Sci., Camb.* **27**, 317–331.

Dick, J. B. & Loader, P. T. (1958). *Temperatures and humidities in pig houses; a study of internal climate in relation to heat and moisture production.* Note E853, Dept. Sci. Industr. Res. Building Research Station, Watford, England.

Donhoffer, Sz., Szegvari, Gy., Jarai, I. & Farkas, M. (1959). Thermoregulatory heat production in the brain. *Nature, Lond.* **184**, 993–994.

Doornebal, H., Asdell, S. A. & Wellington, G. H. (1962). Chromium-51 determined red cell volume as an index of 'lean body mass' in pigs. *J. anim. Sci.* **21**, 461–463.

Dunne, H. W. (1964). *Diseases of swine,* 2nd ed. Iowa State University Press, Ames, Iowa.

Edney, E. B. (1957). *The survival of animals in hot deserts.* Oxford University Press.

Elneil, H. & McCance, R. A. (1965). The effect of environmental temperature on the composition and carbohydrate metabolism of the newborn pig. *J. Physiol., Lond.* **179**, 278–284.

Elsley, F. W. H., Bannerman, M., Bathurst, E. V. J., Bracewell, A. G., Cunningham, J. M. M., Dodsworth, T. L., England, G. J., Forbes, T. J. & Laird, R. (1967) A study of the effects of different levels of feeding in pregnancy and lactation on sow productivity. *Anim. Prod.* **9**, 270–271.

Euler, C. von (1961). Physiology and pharmacology of temperature regulation. *Pharmac. Rev.* **13**, 361–398.

Field, H. I. (1964). *Diseases of pigs. Min. of Agr., Fisheries and Food, Bull.* **171.** London: Her Majesty's Stationery Office.

Filer, L. J., Jr. & Churella, H. (1963). Relationship of body composition, chemical maturation, homeostasis, and diet in the new-born mammal. *Ann. N.Y. Acad. Sci.* **110**, 380–394.

Filer, L. J., Jr., Owen, G. M. & Fomon, S. J. (1966). Effect of age, sex and diet on carcass composition of infant pigs. In *Swine in biomedical research,* eds. L. K. Bustad & R. O. McClellan, pp. 141–150. Battelle Memorial Institute.

Findlay, J. D. (1950). The effects of temperature, humidity, air movement and solar radiation on the behaviour and physiology of cattle and other farm animals. *Hannah Dairy Res. Inst. Bull.* 9.

Findlay, J. D., McLean, J. A. & Bennet, R. D. (1959). A climatic laboratory for farm animals. *Heating & Ventil. Engnr.* Sept.–Oct. 1959.

Folk, G. E., Jr. (1966). *Introduction to environmental physiology.* London: Kimpton.

Fox, M. W. (1967). Influence of domestication upon behaviour of animals. *Vet. Rec.* **80**, 696–702.

Fraps, R. M. (1962). Effects of external factors on the activity of the ovary. In *The Ovary*, ed. S. Zuckerman. New York: Academic Press.

Freeman, B. M. (1966). The effects of cold, noradrenaline and adrenaline upon the oxygen consumption and carbohydrate metabolism of the young fowl (Gallus domesticus). *Comp. Biochem. Physiol.* **18**, 369–382.

Fregly, M. J. & Waters, I. W. (1966). Water intake of rats immediately after exposure to a cold environment. *Can. J. Physiol.* **44**, 651–652.

Fuhrman, G. J. & Fuhrman, F. A. (1959). Oxygen consumption of animals and tissues as a function of temperature. *J. gen. Physiol.* **42**, 715–722.

Fuller, M. F. (1965). The effect of environmental temperature on the nitrogen metabolism and growth of the young pig. *Br. J. Nutr.* **19**, 531–546.

Funk, J. P. (1959). Improved polythene-shielded net radiometer. *J. Sci. Instrum.* **36**, 267–270.

Gadzhiev, G. K. (1962). The effect of active movement on the growth and development of pigs (Russian). *Trud. Novocherkassk Zooteh Vet. Inst.* **14**, 9–21.

Gagge, A. P. (1936). The linerarity criterion as applied to partitional calorimetry. *Am. J. Physiol.* **116**, 656–668.

Gagge, A. P. (1940). Standard operative temperature, a generalized temperature scale, applicable to direct and partitional calorimetry. *Am. J. Physiol.* **131**, 93–103.

Gagge, A. P. (1965). Operative temperature, a physical measure of thermal comfort and thermal equilibrium. *Internat. Physiol. Congr.* Tokyo.

Gagge, A. P., Herrington, L. P. & Winslow, C.-E. A. (1937). Thermal interchanges between the human body and its atmospheric environment. *Am. J. Hyg.* **26**, 84–102.

Gagge, A. P., Stolwijk, J. A. J. & Hardy, J. D. (1965). A novel approach to measurement of man's heat exchange with a complex radiant environment. *Aerospace Medicine* **36**, 432–435.

Gagge, A. P., Winslow, C.-E. A. & Herrington, L. P. (1938). The influence of clothing on the physiological reactions of the human body to varying environmental temperatures. *Am. J. Physiol.* **124**, 30–50.

Galvao, P. E., Tarasantchi, J. & Guertzenstein, P. (1965). Heat production of tropical snakes in relation to body weight and body surface. *Am. J. Physiol.* **209**, 501–506.

Gehlbach, G. D., Becker, D. E., Cox, J. L., Harmon, B. G. & Jensen, A. H. (1966). Effects of floor space allowance and number per group on performance of growing-finishing swine. *J. anim. Sci.* **25,** 386–391.

Gelineo, S. (1954). Le développement ontogénique de la thermorégulation chez le Chien. *C. r. Seanc. Soc. Biol.* **148,** 1483–1485.

Gelineo, S. (1959). The development of homeothermy in mammals. (Russian.) *Usp. Sovrem. biol.* **47,** 108–120.

Giaja, J. (1925). Le métabolisme de sommet et let quotient métabolique. *Annls physiol. physicochem. biol.* **1,** 596–627.

Giaja, J. (1938*a*). L'homéothermie. *Actualités scient. industr.* 576. IX, 1.

Giaja, J. (1938*b*). Le thermorégulation. *Actualités scient. industr.* 577. X, II.

Goodwin, R. F. W. (1957*a*). The relationship between the concentration of blood sugar and some vital body functions in the new-born pig. *J. Physiol., Lond.* **136,** 208–217.

Goodwin, R. F. W. (1957*b*). The concentration of blood sugar during starvation in the new-born calf and foal. *J. comp. Path. Ther.* **67,** 289–296.

Goodwin, R. F. W. (1963). The economic effect of enzootic pneumonia in a large herd of pigs. *Br. Vet. J.* **119,** 298–306.

Gordon, W. A. M. (1963*a*). Environmental studies in pig housing. IV. The bacterial content of air in piggeries and its influence on disease incidence. *Br. Vet. J.* **119,** 263–273.

Gordon, W. A. M. (1963*b*). Environmental studies in pig housing. V. The effects of housing on the degree and incidence of pneumonia in bacon pigs. *Br. Vet. J.* **119,** 307–314.

Graham, N. McC. (1964). Influence of ambient temperature on the heat production of pregnant ewes. *Aust. J. agric. Res.* **15,** 982–988.

Graham, N. McC., Wainman, F. W., Blaxter, K. L. & Armstrong, D. G. (1959). Environmental temperature, energy metabolism and heat regulation in sheep. I. Energy metabolism in closely clipped sheep. *J. agric. Sci., Camb.* **52,** 13–24.

Graham, R., Sampson, J. & Hester, H. R. (1941). Acute hypoglycemia in newly born pigs. (So-called Baby Pig Disease). *Proc. Soc. Exp. Biol. Med.* **47,** 338–339.

Gridgeman, N. T. & Héroux, O. (1965). Relation of oxygen uptake to body weight of rats at different environmental temperatures. *Can. J. Physiol. Pharmacol.* **43,** 351–357.

Hafez, E. S. E., Mauer, R. E. & Ensminger, M. E. (1958). Maternal environment and fetal development in the pig. *Growth* **22,** 269–289.

Hafez, E. S. E., Sumption, L. J. & Jakway, J. S. (1962). The behaviour of swine. In *The behaviour of domestic animals*, pp. 334–369, ed. E. S. E. Hafez. Baillière, Tindall and Cox, London.

Haldane, J. S. (1889). *The methods of investigating quantitatively the heat production and the respiratory exchange of material in animals.* Thesis, University of Edinburgh.

Hammel, H. T. & Hardy, J. D. (1963). A gradient type of calorimeter for measurement of thermoregulatory responses in the dog. In *Temperature, its measurement and control in science and industry.* Vol. 3, Part 3, ed. J. D. Hardy, pp. 31–42. New York: Reinhold.

Hammond, J. (1961). Effect of nutrition on the stage of development of the young at birth in farm animals. In *Somatic stability in the newly born*. Ciba Found. Symp. ed. G. E. W. Wolstenholme & M. O'Connor, pp. 5–15. London: Churchill.

Hanawalt, V. M. & Sampson, J. (1947). Studies on baby pig mortality. V. Relationship between age and time of onset of acute hypoglycemia in fasting new-born pigs. *Am. J. vet. Res.* **8**, 235–243.

Hardy, J. D. (1939). The radiating power of human skin in the infra-red. *Am. J. Physiol.* **127**, 454–462.

Hardy, J. D. (1949). Heat Transfer. In *Physiology of heat regulation and science of clothing*, ed. L. H. Newburgh, pp. 78–108. Philadelphia: Saunders.

Hardy, J. D. (1961). Physiology of temperature regulation. *Physiol. Rev.* **41**, 521–606.

Haring, F., Gruhn, R., Smidt, D. & Scheven, B. (1966). Miniature swine development for laboratory purposes. In *Swine in biomedical research*, eds. L. K. Bustad & R. O. McClellan, pp. 789–796. Battelle Memorial Institute.

Harrington, G. (1962). Progeny testing as an aid to pig improvement. *Outlook on Agriculture*, **3**, 180–189.

Hart, J. S. (1957). Climatic and temperature induced changes in the energetics of homeotherms. *Revue can. Biol.* **16**, 133–174.

Hart, J. S. (1963). Physiological responses to cold in nonhibernating homeotherms. In *Temperature, its measurement and control in science and industry*, Vol. 3, Part 3, ed. J. D. Hardy, pp. 373–406. New York: Reinhold.

Hart, J. S., Héroux, O., Cottle, W. H. & Mills, C. A. (1961). The influence of climate on metabolic and thermal responses of infant caribou. *Can. J. Zool.* **39**, 845–856.

Harvey, C. N. (1959). *A bibliography of farm buildings Research, 1945–1958. Part 1: Buildings for pigs.* Agricultural Research Council, London.

Harvey, C. N. (1962). *A bibliography of farm buildings research. Part 1: Buildings for pigs, 1st supplement, 1958–1961.* Agricultural Research Council, London.

Harvey, C. N. & Easton, P. H. (1965). *A bibliography of farm buildings research. Part 1: Buildings for pigs, 2nd supplement, 1962–1964.* Agricultural Research Council, London.

Harvey, C. N. & Easton, P. H. (1967). *A bibliography of farm buildings research. Supplement for 1965.* Agricultural Research Council, London.

Hashizume, T., Morimoto, H., Hamada, T., Masubuchi, T., Abe, M., Horii, S., Tanaka, K., Zitsukawa, Y., Yokota, C. & Anbo, S. Studies on the feeding standard for Dairy cattle. III. The influence of the environmental temperature on the maintenance requirements of dairy cattle. *Spec. Rep. of Nat. Inst. of Animal Indust.* No. 2, March, 1964, Chiba, Japan.

Hatfield, H. S. & Wilkins, F. J. (1950). A new heat-flow meter. *J. sci. Instrum.* **27**, 1–3.

Hayward, J. S. (1965). Microclimate temperature and its adaptive significance in six geographic races of Peromysus. *Can. J. Zool.* **43**, 341–350.

Heitman, H. Jr., Hahn, L., Bond, T. E. & Kelly, C. F. (1962). The effects of modified summer environment on swine behaviour. *Anim. Behaviour* **10**, 15–19.

Heitman. H. Jr. & Hughes, E. H. (1949). The effects of air temperature and relative humidity on the physiological well being of swine. *J. anim. Sci.* **8**, 171–181.

Heitman, H., Jr., Kelly, C. F. & Bond, T. E. (1958). Ambient air temperature and weight gain in swine. *J. anim. Sci.* **17**, 62–67.

Heitman, H., Jr., Kelly, C. F. & Hughes, E. H. (1949). California psychrometric chamber for livestock environmental studies. *J. Anim. Sci.* **8**, 459–463.

Hemmingsen, A. M. (1960). Energy metabolism as related to body size and respiratory surfaces, and its evolution. *Rep. Steno Hosp., Copenhagen* **9**, part 2, 7–110.

Henschel, A. (1964). Laboratory facilities for adaptation research: high and low temperatures. In *Adaptation to environment*, ed. D. B. Dill, pp. 323–328. American Physiological Society.

Hentschel, G. (1964). Sports and climate. In *Medical Climatology*, ed. S. Licht. Newhaven, Connecticut: E. Licht.

Hess, E. A. & Bailey, C. B. M. (1961). Comparative physiological effects of cold on farm and laboratory animals. *Anim. Breeding Abstr.* **29**, 379–392.

Hey, E. N. & Mount, L. E. (1966). Temperature control in incubators. *Lancet* **ii**, 202–203.

Hey, E. N. & Mount, L. E. (1967). Heat losses from babies in incubators. *Arch. dis. Childh.* **42**, 75–84.

Hill, A. V. & Hill, A. M. (1914). A self-recording calorimeter for large animals. *J. Physiol., Lond.* **48**, xiii.

Hill, June R. (1961). Reaction of the new-born animal to environmental temperature. *Br. med. Bull.* **17**, 164–167.

Hill, June R. & Rahimtulla, K. A. (1965). Heat balance and the metabolic rate of new-born babies in relation to environmental temperature; and the effect of age and of weight on basal metabolic rate. *J. Physiol., Lond.* **180**, 239–265.

Holmes, C. W. (1966). *Studies on the effects of environment on heat losses from pigs*. Ph.D. Thesis, The Queen's University of Belfast.

Holmes, C. W. (1968). Heat losses from young pigs at three environmental temperatures, measured in a direct calorimeter. *Anim. Prod.*, **10**, 135–147.

Holmes, C. W. & Mount, L. E. (1967). Heat loss from groups of growing pigs under various conditions of environmental temperature and air movement. *Anim. Prod.* **9**, 435–452.

Holmes, R. M. & Dingle, A. N. (1965). The relationship between the macro- and micro-climate. *Agr. Meteorol.* **2**, 127–133.

Holub, A., Forman, Z. & Ježková, D. (1957). Development of chemical thermoregulation in piglets. *Nature, Lond.* **180**, 858–859.

Howie, J. W. (1951). Preventing losses among young pigs. *Scottish Agriculture* **30**, 188–191.

Hull, D. (1966). The structure and function of brown adipose tissue. *Br. med. Bull.* **22,** 92–96.

Hull, D. & Segall, M. M. (1964). The effect of removing brown adipose tissue on heat production in the new-born rabbit. *J. Physiol., Lond.* **175,** 58P.

Inglestedt, S. & Toremalm, N. G. (1961). Air flow patterns and heat transfer within the respiratory tract. *Acta physiol. scand.* **51,** 204–217.

Inglis, J. S. S. & Robertson, A. (1949). Hygienic aspects of pig housing: a review. *Vet. Rec.* **61,** 141–145.

Inglis, J. S. S. & Robertson, A. (1953). The measurement of heat loss through floors. *Vet. Rec.* **65,** 875–876.

Ingram, D. L. (1964a). The effect of environmental temperature on body temperatures, respiratory frequency and pulse rate in the young pig. *Res. vet. Sci.* **5,** 348–356.

Ingram, D. L. (1964b). The effect of environmental temperature on heat loss and thermal insulation in the young pig. *Res. vet. Sci.* **5,** 357–364.

Ingram, D. L. (1965a). Evaporative cooling in the pig. *Nature, Lond.* **207,** 415–416.

Ingram, D. L. (1965b). The effect of humidity on temperature regulation and cutaneous water loss in the young pig. *Res. vet. Sci.* **6,** 9–17.

Ingram, D. L. (1967). Stimulation of cutaneous glands in the pig. *J. comp. Path.* **77,** 93–98.

Ingram, D. L. & Mount, L. E. (1965). The metabolic rates of young pigs living at high ambient temperatures. *Res. vet. Sci.* **6,** 300–306.

Ingram, D. L. & Slebodzinski, A. (1965). Oxygen consumption and thyroid gland activity during adaptation to high ambient temperatures in young pigs. *Res. vet. Sci.* **6,** 522–530.

Ingram, D. L. & Whittow, G. C. (1962). The effects of variations in respiratory activity and in the skin temperatures of the ears on the temperature of the blood in the external jugular vein of the ox (Bos taurus). *J. Physiol., Lond.* **163,** 211–221.

Irving, L. (1956a). Physiological insulation of swine as bare-skinned mammals. *J. appl. Physiol.* **9,** 414–420.

Irving, L. (1956b). Metabolism and insulation of swine as bare-skinned mammals. *J. appl. Physiol.* **9,** 421–426.

Irving, L. (1964). Terrestrial animals in cold: birds and mammals. In *Adaptation to Envt.*, ed. D. B. Dill, pp. 361–377. American Physiological Society.

Jakob, M. (1949). *Heat transfer.* Vol. I. New York: Wiley.

Jakob, M. (1957). *Heat transfer.* Vol. II. New York: Wiley.

Jenkinson, D. M. & Blackburn, P. S. (1967). The distribution of nerves, monoamine oxidase and cholinesterase in the skin of the pig. *Res. vet. Sci.* **8,** 306–312.

Joblin, A. D. H. (1966). The estimation of carcass composition in bacon weight pigs. 1. The use of ultrasonic and carcass measurements. *New Zealand J. agr. Res.* **9,** 108–126.

Jones, H. W., Pickett, R. A., Kadlec, J. E. & Morris, W. H. (1965). A winter comparison of bedding vs. no bedding in an open front growing-finishing house. *Purdue Univ. Agric. Expt. Sta. Res. Prog. Rep.* 206.

Joyce, J. P. & Blaxter, K. L. (1964). The effect of air movement, air temperature and infra-red radiation on the energy requirements of sheep. *Br. J. Nutr.* **18**, 5–27.

Joyce, J. P., Blaxter, K. L. & Park, C. (1966). The effect of natural outdoor environments on the energy requirements of sheep. *Res. vet. Sci.* **7**, 342–359.

Kaemmerer, K. (1966). Fundamental studies on miniature pigs (German). (Abstr.). *Biol. Abstr.* **47**, 73238.

Kare, M. R., Pond, W. C. & Campbell, J. (1965). Observations on the taste reactions in pigs. *Anim. Behav.* **13**, Suppl. 1, 265–269.

Karlberg, P., Moore, R. E. & Oliver, T. K. (1962). The thermogenic response of the new-born infant to noradrenaline. *Acta paediat., Stockh.* **51**, 284–292.

Kay, M. & Jones, A. S. (1964). The relationship between body density and body fat for living pigs. *Anim. Prod.* **6**, 258.

Kelly, C. F., Bond, T. E. & Garrett, W. (1964). Heat transfer from swine to a cold slab. *Trans. Am. Soc. agric. Engrs* **7**, 34–35, 37.

Kelly, C. F., Bond, T. E. & Heitman, H., Jr. (1954). The role of thermal radiation in animal ecology. *Ecology* **35**, 562–569.

Kelly, C. F., Bond, T. E. & Heitman, H., Jr. (1963). Direct 'air' calorimetry for livestock. *Trans. Am. Soc. agric. Engrs* **6**, 126–128.

Kemény, A., Gáspár, Z. N., Pethes, G., Tóth, B. & László, D. (1955). Studies on the carbohydrate metabolism in fasting new-born and sucking pigs. *Acta Vet. Acad. Sci. Hung.* **5**, 115–131.

Kerslake, D. McK. (1963). Errors arising from the use of mean heat exchange coefficients in the calculation of the heat exchanges of a cylindrical body in a transverse wind. In *Temperature, its measurement and control in science and industry*, Vol. 3, Part 3, ed. J. D. Hardy, pp. 183–190. New York: Reinhold.

Kidder, D. E., Manners, M. J. & McCrea, M. R. (1963). The digestion of sucrose by the piglet. *Res. vet. Sci.* **4**, 131–144.

Kidder, D. E., Manners, M. J., McCrea, M. R. & Weaver, B. M. Q. (1963). Fructose utilization in the piglet. *Res. vet. Sci.* **4**, 145–150.

Kleiber, M. (1932). Body size and metabolism. *Hilgardia*, **6**, 315–353.

Kleiber, M. (1947). Body size and metabolic rate. *Physiol. Rev.* **27**, 511–541.

Kleiber, M. (1961). *The fire of life.* New York: Wiley.

Kleiber, M. (1965). Metabolic body size. In *Energy Metabolism*, ed. K. L. Blaxter, pp. 427–435. London: Academic Press.

Kleiber, M. (1967). Further consideration of the relation between metabolic rate and body size. European Association for Animal Production, *4th Symposium on Energy Metabolism*.

Kleiber, M. & Winchester, C. (1933). Temperature regulation in baby chicks. *Proc. Soc. exp. Biol. Med.* **31**, 158–159.

Kločkova, A. J. (1961). The effect of photoperiodicity on sexual function of sows. Quoted in *Anim. Breeding Abstr.* **29**, 2255.

Kločkova, A. J. & Emme, A. M. (1964). The effect of photoperiodicity on pigs. Quoted in *Anim. Breeding Abstr.* **32**, 338.

Klopfer, F. D. (1966). Visual learning in swine. In *Swine in Biomedical Research*, pp. 559–574, eds. L. K. Bustad & R. O. McClellan. Battelle Memorial Institute.

Koch, W. (1962). Relationship between air temperature and mean radiant temperature in thermal comfort. *Nature, Lond.* **196**, 587.

Konyukhova, V. O. (1966). Conditional reflexes to speech stimuli in swine. (Russian.) Abstr. in *Biol. Abstr.* **47**, 100583.

Kuppenheim, H. F., Dimitroff, J. M., Melotti, P. M., Graham, I. C. & Swanson, D. W. (1956). *J. appl. Physiol.* **9**, 75–78.

Ladrat, J. (1964). Exercise musculaire et production de viande. *Rev. de Méd. Vét.* **115**, 62–64.

Landsberg, H. E. (1964). Controlled climate (outdoor and indoor). In *Medical Climatology*, ed. S. Licht. New Haven, Connecticut: E. Licht.

Lasiewski, R. C. (1963). Oxygen consumption of torpid, resting, active and flying hummingbirds. *Physiol. Zoöl.* **36**, 122–140.

Lasiewski, R. C., Acosta, A. L. & Bernstein, M. H. (1966). Evaporative water loss in birds. 1. Characteristics of the open flow method of determination, and their relation to estimates of thermoregulatory ability. *Comp. Biochem. Physiol.* **19**, 445–457.

Lavoisier, A. L. & Laplace, de. (1780). Mémoire sur la chaleur. *Histoire de l'Acad. Roy. des Sci.* **85**, 355.

LeBlanc, J. & Mount, L. E. (1968). The effects of noradrenaline and adrenaline on oxygen consumption rate and arterial blood pressure in the new-born pig. *Nature, Lond.* **217**, 77–78.

Lee, D. H. K. & Vaughan, J. A. (1964). Temperature equivalent of solar radiation on man. *Int. J. Biometeor.* **8**, 61–69.

Lentz, C. P. & Hart, J. S. (1960). The effect of wind and moisture on heat loss through the fur of new-born caribou. *Can. J. Zool.* **38**, 679–688.

Lewis, T. (1930). Observations upon the reactions of vessels of human skin to cold. *Heart.* **15**, 177–208.

Lucas. I. A. M. (1964). Modern methods of pig nutrition: the impact of recent research. *Vet. Rec.* **76**, 101–109.

Ludvigsen, J. & Thorbek, G. (1955). Heat production of pigs of various ages. *Beretu. f. forsogslat.* Copenhagen. 283.

Lusk, G. (1928). *The elements of the science of nutrition*, 4th edition. Philadelphia: Saunders.

McBride, G. (1963). The 'teat order' and communication in young pigs. *Anim. Behav.* **11**, 53–56.

McBride, G., James, J. W. & Wyeth, G. S. F. (1965). Social behaviour of domestic animals. VII. Variation in weaning weight in pigs. *Anim. Prod.* **7**, 67–74.

McCance, R. A. (1960). Severe undernutrition in growing and adult animals. I. Production and general effects. *Br. J. Nutr.* **14**, 59–73.

McCance, R. A. & Mount, L. E. (1960). Severe undernutrition in growing and adult animals. 5. Metabolic rate and body temperature in the pig. *Br. J. Nutr.* **14**, 509–518.

McCance, R. A. & Windowson, E. M. (1959). The effect of lowering the ambient temperature on the metabolism of the new-born pig. *J. Physiol., Lond.* **147**, 124–134.

McGuire, J. H. (1953). *D.S.I.R. Fire Research Special Rep. No. 2.* London, H.M.S.O.

MacHardy, F. W. (1960). Transient heat flow in piggery floors. *Can. agric. Engng.* **2**, 13–15.

McIntyre, D. G. & Ederstrom, H. E. (1958). Metabolic factors in the development of homeothermy in dogs. *Am. J. Physiol.* **194**, 293–296.

McLagan, J. R. & Thomson, W. (1950). Effective temperature as a measure of environmental conditions for pigs. *J. agric. Sci., Camb* **40**, 367–374.

McLean, J. A. (1963). The partition of insensible losses of body weight and heat from cattle under various climatic conditions. *J. Physiol., Lond.* **167**, 427–447.

Mann, T. P. & Elliott, R. I. K. (1957). Neonatal cold injury due to accidental exposure to cold. *Lancet*, **i**, 229–234.

Meeh, K. (1879). Oberflächenmessungen des menschlichen Körpers. *Z. Biol.* **15**, 425–458.

Meissl, E. (1886). Untersuchungen über den Stoffwechsel des Schweines, *Z. Biol.* **22**, 63–160.

Miller, A. T., Jr. & Blyth, C. S. (1958). Lack of insulating effect of body fat during exposure to internal and external heat loads. *J. appl. Physiol.* **12**, 17–19.

Molnar, G. W. & Rosenbaum, J. C., Jr. (1963). Surface temperature measurement with thermocouples. In *Temperature, its measurement and control in science and industry*, Vol. 3, Part 3, ed. J. D. Hardy, pp. 3–11. New York: Reinhold.

Montagna, W. (1966). The microscopic anatomy of the skin of swine and man. In *Swine in biomedical research*, eds. L. K. Bustad & R. O. McClellan, pp. 285–286. Battelle Memorial Institute.

Moore, R. E. & Underwood, Mary C. (1960). Noradrenaline as a possible regulator of heat production in the new-born kitten. *J. Physiol., Lond.* **150**, 13–14 P.

Moore, R. E. & Underwood, Mary C. (1963). The thermogenic effects of noradrenaline in new-born and infant kittens and other small mammals. A possible hormonal mechanism in the control of heat production. *J. Physiol., Lond.* **168**, 290–317.

Morrill, C. C. (1952a). Studies on baby pig mortality. VIII. Chemical observations on the new-born pig, with special reference to hypoglycemia. *Am. J. vet. Res.* **13**, 164–170.

Morrill, C. C. (1952b). Studies on baby pig mortality. IX. Some morphological observations on the new-born pig, with special reference to hypoglycemia. *Am. J. vet. Res.* **13**, 171–180.

Morrill, C. C. (1952c). Studies on baby pig mortality. X. Influence of environmental temperature on fasting new-born pigs. *Am. J. vet Res.* **13**, 322–324.

Morrill, C. C. (1952d). Studies on baby pig mortality. XI. A note on the influence of fasting on body temperature, body weight, and liver weight of the new-born pig. *Am. J. vet. Res.* **13**, 325–326.

Morris, D. (1965). *The Mammals*. London: Hodder and Stoughton.

Morrison, S. R., Bond, T. E.. & Heitman, H. (1967). Skin and lung moisture loss from swine. *Trans. Am. Soc. agric. Engrs* **10**, 691–692, 696.

Morrison, S. R., Bond, T. E. & Heitman, H. (1967). Effect of humidity on swine at high temperature. Paper No. 67–420 (Mimeograph). Am. Soc. agric. Engrs, St. Joseph, Michigan.

Morrison, S. R., Heitman, H., Bond, T. E. & Finn-Kelcey, P. (1966). The influence of humidity on growth rate and feed utilization of swine. *Proc. 4th Internat. Biometeorol. Congr.*, New Jersey, U.S.A.

Mount, L. E. (1958). Change in oxygen consumption of the new-born pig with a fall in environmental temperature. *Nature, Lond.* **182**, 536–537.

Mount, L. E. (1959). The metabolic rate of the new-born pig in relation to environmental temperature and to age. *J. Physiol., Lond.* **147**, 333–345.

Mount, L. E. (1960). The influence of huddling and body size on the metabolic rate of the young pig. *J. agric. Sci., Camb.* **55**, 101–105.

Mount, L. E. (1961). Metabolic rate and body temperature in the newly born. In *Somatic stability in the newly born*, Ciba Found. Symp., eds. G. E. W. Wolstenholme and M. O'Connor, pp. 117–130. London: Churchill.,

Mount, L. E. (1962a). Evaporative heat loss in the new-born pig. *J. Physiol., Lond.* **164**, 274–281.

Mount, L. E. (1962b). Developmental aspects of the environmental physiology of the pig. In *Nutrition of pigs and poultry*, eds. J. T. Morgan and D. Lewis, pp. 77–86. London: Butterworths.

Mount, L. E. (1963a). Responses to thermal environment in new-born pigs. *Fedn Proc. Fedn Am. Socs exp. Biol.* **22**, 818–823.

Mount, L. E. (1963b). The thermal insulation of the new-born pig. *J. Physiol., Lond.* **168**, 698–705.

Mount, L. E. (1963c). The environmental temperature preferred by the young pig. *Nature, Lond.* **199**, 122–123.

Mount, L. E. (1964a). The tissue and air components of thermal insulation in the new-born pig. *J. Physiol., Lond.* **170**, 286–295.

Mount, L. E. (1964b). Radiant and convective heat loss from the new-born pig. *J. Physiol., Lond.* **173**, 96–113.

Mount, L. E. (1964c). A simple instrument for the measurement of low air speeds, with special reference to pig housing. *J. agric. Sci., Camb.* **63**, 335–339.

Mount, L. E. (1965). The young pig and its physical environment. In *Energy Metabolism*, ed. K. L. Blaxter, pp. 379–386. London: Academic Press.

Mount, L. E. (1966a). Basis of heat regulation in homeotherms. *Br. med. Bull.* **22**, 84–87.

Mount, L. E. (1966b). Thermal and metabolic comparisons between the new-born pig and human infant. In *Swine in biomedical research*, eds. L. K. Bustad & R. O. McClellan, pp. 501–509. Battelle Memorial Institute.

Mount, L. E. (1966c). The effect of wind-speed on heat production in the new-born pig. *Q. Jl exp. Physiol.* **51**, 18–26.

Mount, L. E. (1966d). The Babraham Pig Section. In *Swine in biomedical research*, eds. L. K. Bustad & R. O. McClellan, pp. 633–636. Battelle Memorial Institute.

Mount, L. E. (1967a). The heat loss from new-born pigs to the floor. *Res. vet. Sci.* **8**, 175–186.

Mount, L. E. (1967b). Adaptation of swine. In *Adaptation of domestic animals*, ed. E. S. E. Hafez. Philadelphia: Lea and Febiger.

Mount, L. E., Holmes, C. W., Start, I. B. & Legge, A. J. (1967). A direct calorimeter for the continuous recording of heat loss from groups of growing pigs over long periods. *J. agric. Sci., Camb.* **68**, 47–55.

Mount, L. E. & Ingram, D. L. (1965). The effects of ambient temperature and air movement on localized sensible heat-loss from the pig. *Res. vet. Sci.* **6**, 84–91.

Mount, L. E., Lister, D. & McCance, R. A. (1963). Severe undernutrition in growing and adult animals. 11. The first effects of rehabilitation on the metabolic rate and body temperature. *Br. J. Nutr.* **17**, 403–413.

Mount, L. E. & Rowell, J. G. (1960a). Body weight and age in relation to the metabolic rate of the young pig. *Nature, Lond.* **186**, 1054–1055.

Mount, L. E. & Rowell, J. G. (1960b). Body size, body temperature and age in relation to the metabolic rate of the pig in the first five weeks after birth. *J. Physiol., Lond.* **154**, 408–416.

Mount, L. E. & Willmott, Jane V. (1967). The relation between spontaneous activity, metabolic rate and the 24-hour cycle in mice at different environmental temperatures. *J. Physiol., Lond.* **190**, 371–380.

Neergaard, L. & Thorbek, G. (1967). Variation in heat production in growing pigs. 1. Some observations on the relationship between feed intake and heat production in pigs fed barley and skim-milk. European Association for Animal Production *4th Symposium on Energy Metabolism*.

Newland, H. W., McMillen, W. N. & Reineke, E. P. (1952). Temperature adaptation in the baby pig. *J. anim. Sci.* **11**, 118–133.

Newton, W. C. & Sampson, J. (1951). Studies on baby pig mortality. VII. The effectiveness of certain sugars other than glucose in alleviating hypoglycemic coma in fasting new-born pigs. *Cornell Vet.* **41**, 377–381.

Norman, J. N. (1965). Cold exposure and patterns of activity at a polar station. *Br. Antarct. Surv. Bull.* No. 6, 1–13.

Oslage, H. J. & Fliegel, H. (1965). Nitrogen and energy metabolism of growing-fattening pigs with an approximately maximal feed intake. In *Energy Metabolism*, ed. K. L. Blaxter, pp. 297–308. London: Academic Press.

Parry, M. (1957). Local climates and house heating. *Adv. Sci.* **13**, 326–331.

Pearson, O. P. (1948). Metabolism of small mammals, with remarks on the lower limit of mammalian size. *Science,* **108**, 44.

Pechert, von H. & Cermak, H. (1958). Versuche mit verschiedenen Fussboden belägen in Schweineställen (1). *Arch. f. Tierzucht.* **1,** 257–276.

Perry, J. S. (1956). Observations on reproduction in a pedigree herd of Large White pigs. *J. agric. Sci., Camb.* **47,** 332–343.

Perry, J. S. & Rowlands, I. W. (1962). Early pregnancy in the pig. *J. Reprod. Fertil.* **4,** 175–188.

Pomeroy, R. W. (1953). Studies on piglet mortality. 1. Effect of low temperature and low plane of nutrition on the rectal temperature of the young pig. *J. agric. Sci., Camb.* **43,** 182–191.

Pomeroy, R. W. (1960a). Infertility and neonatal mortality in the sow. I. Lifetime performance and reasons for disposal of sows. *J. agric. Sci., Camb.* **54,** 1–17.

Pomeroy, R. W. (1960b). Infertility and neonatal mortality in the sow. II. Experimental observations on sterility. *J. agric. Sci., Camb.* **54,** 18–30.

Pomeroy, R. W. (1960c). Infertility and neonatal mortality in the sow. III. Neonatal mortality and foetal development. *J. agric. Sci., Camb.* **54,** 31–56.

Pomeroy, R. W. (1960d). Infertility and neonatal mortality in the sow. IV. Further observations and conclusions. *J. agric. Sci., Camb.* **54,** 57–66.

Pond, W. G., Barnes, R. J., Reid, I., Krook, L., Kwong, E. & Moore, A. U. (1966). Protein deficiency in baby pigs. In *Swine in Biomedical Research*, eds. L. K. Bustad & R. O. McClellan, pp. 213–224. Battelle Memorial Institute.

Pope, C. H. (1962). On the natural superiority of pigs. *Turtox News* **40,** 190–193, 230–231, 258–261.

Prosser, C. L. (1950). Temperature: metabolic aspects and perception. In *Comparative Animal Physiology*, ed. C. L. Prosser, pp. 341–380. Philadelphia: Saunders.

Prosser, C. L. (1964). In *Adaptation to environment*, ed. D. B. Dill. American Physiological Society.

Prouty, L. R., Barrett, M. J. & Hardy, J. D. (1949). A simple calorimeter for the simultaneous determination of heat loss and heat production in animals. *Rev. Sci. Instrum.* **20,** 357–363.

Provins, K. A., Hellon, R. F., Bell, C. R. & Hirons, R. W. (1962). Tolerance to heat of subjects engaged in sedentary work. *Ergonomics* **5,** 93–97.

Prychodko, W. (1958). The effect of aggregation of laboratory mice on food intake at different temperatures. *Ecology,* **39,** 500–503.

Pullar, J. D. (1958). Direct calorimetery of animals by the gradient layer principle. *First Symposium on Energy Metabolism*, European Association for Animal Production, Rome.

Pullar, J. D. (1963). Energy metabolism, in *Progress in Nutrition and Allied Sciences*, ed. D. P. Cuthbertson, pp. 187–197. Edinburgh: Oliver and Boyd.

Rathnasabapathy, V., Lasley, J. F. & Mayer, D. T. (1956). Genetic and environmental factors affecting litter size in swine. *Missouri Agric. Expt. Sta. Res. Bull.* 615.

Reignault, V. & Reiset, J. (1849). Recherches chimiques sur la respiration des animaux. *Ann. de chim. et de phys.* Ser. 3, **26**, 299–519.

Rhoad, A. O. (1940). Absorption and reflection of solar radiation in relation to coat colour in cattle. *Proc. Am. Soc. Anim. Prod.* **33**, 291–293.

Richet, C. (1889). *La chaleur animale.* Bibliothèque Scientifique International, Paris, Felix Alcan.

Ritzman, E. G. & Colovos, N. F. (1941). The dynamic stimulus of food to the pig. *Univ. New Hamps. Tech. Bull.* 75.

Robertson, J. B., Laird, R., Hall, J. K. S., Forsyth, R. J., Thomson, J. M. & Walker-Love, J. (1966). A comparison of two indoor farrowing systems for sows. *Anim. Prod.* **8**, 171–178.

Robinson, K. & Lee, D. H. K. (1941). Reactions of the pig to hot atmospheres. *Proc. Roy. Soc., Queensland* **53**, 145–158.

Roy, J. H. B., Huffman, C. F. & Reineke, E. P. (1957). The basal metabolism of the new-born calf. *Br. J. Nutr.* **11**, 373–381.

Rubner, M. (1894). Die Quelle der thierischen Wärme. *Biol.* **30**, 73–142.

Rubner, M. (1902). *Die Gesetze des Energieverbrauchs bei der Ernährung.* Leipzig und Wien: Deutiche.

Sainsbury, D. W. B. (1954). *Studies on the environment in pig houses.* Ph.D. Thesis, University of London.

Sainsbury, D. W. (1963). *Pig Housing.* Ipswich, England: Rodale.

Sainsbury, D. W. (1967). *Animal health and housing.* London: Baillière, Tindall and Cassell.

Sampson, J., Taylor, R. B. & Smith, J. C. (1955). Hypoglycemic coma and convulsions in fasting baby lambs. *Cornell Vet.* **45**, 10–15.

Sarrus & Rameaux (1839). Mémoire addressé à l'Académie Royale. *Bulletin de l'académie royale de médecine,* **3**, 1094–1100.

Schmidt-Nielsen, K. S. (1964). *Desert animals; physiological problems of heat and water.* Oxford University Press.

Scholander, P. F. (1958). Counter current exchange; a principle in biology. *Hvalrådets Skrifter,* No. 44. Oslo.

Scholander, P. F., Hock, R., Walters, V. & Irving, L. (1950). Adaptation to cold in arctic and tropical mammals and birds in relation to body temperature, insulation and basal metabolic rate. *Biol. Bull., Wood's Hole,* **99**, 259–271.

Shanklin, M. D. & Stewart, R. E. (1958). Relief of thermally induced stress in dairy cattle by radiation cooling. *Univ. Miss Agr. Expt. Sta. Res. Bull.* 670.

Shelley, Heather J. (1961). Glycogen reserves and their changes at birth and in anoxia. *Br. med. Bull.* **17**, 137–143.

Sillar, F. C. & Meyler, R. M. (1961). *The Symbolic Pig.* Edinburgh: Oliver and Boyd.

Silverman, W. A. & Agate, F. J. (1964). Variation in cold resistance among small new-born infants. *Biol. Neonat.* **6**, 113–127.

Skinner, B. F. (1938). *The behaviour of organisms; an experimental analysis.* New York: Appleton-Century.

Slebodzinski, A. (1965a). Interaction between thyroid hormone and thyroxine-binding proteins in the early neonatal period. *J. Endocr.* **32**, 45–57.

Slebodzinski, A. (1965*b*). Physiological significance of thyroxine-binding by proteins; the relationship between changes in thyroxine-binding globulin capacity and the daily utilization rate of thyroxine in the new-born pig. *J. Endocr.* **32**, 65–75.

Slebodzinski, A. (1965*c*). Thyroxine utilization rate and some other indices of peripheral iodine metabolism in sucking pigs. *Res. vet. Sci.* **6**, 307–315.

Slee, J. & Sykes, A. R. (1967). Acclimatisation of Scottish Blackface sheep to cold. *Anim. Prod.* **9**, 333–347.

Smith, C. V. (1964). A quantitative relationship between environment comfort and animal productivity. *Agr. Meteorol.*, **1**, 249–270.

Smith, R. E. (1964). Thermoregulatory and adaptive behaviour of brown adipose tissue. *Science*, **146**, 1686–1689.

Sørensen, P. H. (1962). Influence of climatic environment on pig performance. In *Nutrition of pigs and poultry*, eds. J. T. Morgan & D. Lewis, pp. 88–103. Butterworths, London.

Spaendonck, R. L. van & Vanschoubroek, F. X. (1964). Determination of the milk yield of sows and correction for loss of weight due to metabolic processes of piglets during suckling. *Anim. Prod.* **6**, 119–123.

Steen, I. & Steen, J. B. (1965*a*). The importance of the legs in the thermo-regulation of birds. *Acta physiol. scand.* **63**, 285–291.

Steen, I. & Sheen, J. B. (1965*b*). Thermoregulatory importance of the beaver's tail. *Comp. Biochem. Physiol.* **15**, 267–270.

Stewart, R. E. (1953). Absorption of solar radiation by the hair of cattle. *Agric. Engng, St. Joseph, Mich.* **34**, 235–238.

Tangl, F. (1912). Die minimale Erhaltungsarbeit des Schweines (Stoff- and Energieumsats in Hunger). *Biochem. Z.* **44**, 252–278.

Taylor, P. M. (1960). Oxygen consumption in new-born rats. *J. Physiol., Lond.* **154**, 153–168.

Thauer, R. (1961). Mécanismes périphérique et centraux de la régulation de la température. *Archs Sci. physiol.* **15**, 95–123.

Thomson, M. L. (1955). Relative efficiency of pigment and horny layer thickness in protecting the skin of Europeans and Africans against solar ultraviolet radiation. *J. Physiol., Lond.* **127**, 236–246.

Thorbek, G. (1967). Studies on energy metabolism of growing pigs. European Association for Animal Production, *4th Symp. Energy Metabolism.*

Thorbek, G. & Neergaard, L. (1965). Some remarks about instruments and methods applied to an open-circuit respiration apparatus for pigs. In *Energy Metabolism*, ed. K. L. Blaxter, pp. 179–188. London: Academic Press.

Tonks, H. M. & Smith, W. C. (1965). The effect of a high-temperature, high-humidity indoor environment on the performance of bacon pigs. *Anim. Prod.* **7**, 280.

Tregear, R. T. (1965). Hair density, wind speed and heat loss in mammals. *J. appl. Physiol.* **20**, 796–801.

Trow-Smith, R. (1957). *A history of British livestock husbandry to 1700.* London: Routledge and Kegan Paul.

Valen, E. (1958). Oxygen consumption in relation to temperature in some poikilotherms. *Acta physiol. scand.* **42,** 358–362.

Vere, D. W. (1958). Heat transfer measurement in living skin. *J. Physiol., Lond.* **140,** 359–380.

Veterinary Investigation Service (1960). A study of the incidence and causes of mortality in pigs. II. Findings of post mortem examination of pigs. *Vet. Rec.* **72,** 1240–1247.

Walker, D. M. (1959). The development of the digestive system of the young animal. *J. agric. Sci., Camb.* **52,** 357–363.

Walker, J. E. C., Wells, R. E., Jr. & Merrill, E. W. (1961). Heat and water exchange in the respiratory tract. *Am. J. Med.* **30,** 259–267.

Wallace, L. R. (1948). The growth of lambs before and after birth in relation to the level of nutrition. *J. agric. Sci., Camb.* **38,** 93–153, 243–302, 367–401.

Weaver, M. E. & McKean, C. F. (1965). Miniature swine as laboratory animals. *Lab. Animal Care* **15,** 49–56.

Weiner, J. S. (1966). Aspects of adaptation. *Advmt Sci., Lond.* **23,** 370–378.

Weinstein, G. D. (1966). Comparison of turnover time and of keratinous protein fractions in swine and human epidermis. In *Swine in Biomedical Research*, ed. L. K. Bustad & R. O. McClellan, pp. 287–297. Battelle Memorial Institute.

Weiss, B. (1957a). Thermal behaviour of the subnourished and pantothenic-acid-deprived-rat. *J. Comp. Physiol. Psychol.* **50,** 481–485.

Weiss, B. (1957b). Pantothenic acid deprivation and thermal behaviour. of the rat. *Am. J. Clin. Nutr.* **5,** 125–128.

Weiss, B. & Laties, V. G. (1961). Behavioural thermoregulation. *Science* **133,** 1338–1344.

Whittlestone, P. (1960). Virus pneumonia of pigs. *Outlook on Agriculture.* **2,** 284–290.

Widdowson, E. M. (1950). Chemical composition of newly born mammals. *Nature, Lond.* **166,** 626–628.

Winslow, C.-E. A. & Gagge, A. P. (1941). Influence of physical work on physiological reactions to the thermal environment. *Am. J. Physiol.* **134,** 664–681.

Winslow, C.-E. A., Gagge, A. P. & Herrington, L. P. (1939). The influence of air movement upon heat losses from the clothed human body. *Am. J. Physiol.* **127,** 505–518.

Winslow, C.-E. A., Gagge, A. P. & Herrington, L. P. (1940). Heat exchange and regulation in radiant environments above and below air temperature. *Am. J. Physiol.* **131,** 79–92.

Winslow, C.-E. A. & Herrington, L. P. (1949). *Temperature and human life.* Princeton University Press, Princeton, New Jersey.

Winslow, C.-E. A., Herrington, L. P. & Gagge, A. P. (1936a). A new method of partitional calorimetry. *Am. J. Physiol.* **116,** 641–655.

Winslow, C.-E. A., Herrington, L. P. & Gagge, A. P. (1936b). The determination of radiation and convection exchanges by partitional calorimetry. *Am. J. Physiol.* **116,** 669–684.

Winslow, C.-E. A., Herrington, L. P. & Gagge, A. P. (1937*a*). Physiological reactions of the human body to varying environmental temperatures. *Am. J. Physiol.* **120**, 1–22.

Winslow, C.-E. A., Herrington, L. P. & Gagge, A. P. (1937*b*). Relations between atmospheric conditions, physiological reactions and sensations of pleasantness. *Am. J. Hyg.* **26**, 103–115.

Winslow, C.-E. A., Herrington, L. P. & Gagge, A. P. (1938). The relative influence of radiation and convection upon the temperature regulation of the clothed body. *Am. J. Physiol.* **124**, 51–61.

Wolff, H. S. (1958). A knitted wire fabric for measuring mean skin temperature or for body heating. *J. Physiol., Lond.* **142**, 1–2P.

Wood, A. J. & Groves, T. D. D. (1965). Body composition studies on the suckling pig. I. Moisture, chemical fat, total protein, and total ash in relation to age and body weight. *Can. J. anim. Sci.* **45**, 8–13.

Woolsey, C. N. & Fairman, D. (1946). Contralateral, ipsilateral, and bilateral representation of cutaneous receptors in somatic areas I and II of the cerebral cortex of pig, sheep and other mammals. *Surgery*, **19**, 684–702.

Yaglou, C. P. (1927). Temperature, humidity and air movement in industries. The effective temperature index. *J. Indust. Hyg.* **9**, 297–309.

Yaglou, C. P. (1949). Indices of comfort. In *Physiology of heat regulation and the science of clothing*, ed. L. H. Newburgh. Philadelphia: Saunders.

Yaglou, C. P. & Miller, W. E. (1925). Effective temperature with clothing. *Trans. Am. Soc. Heat. Vent. Engrs* **31**, 89–99.

Zausch, M. (1965). Modifications in techniques and apparatus for studies of respiration in small animals. In *Energy Metabolism*, ed. K. L. Blaxter, pp. 159–164. London: Academic Press.

Zhmurin, L. M. (1959). Gas and energy exchange in piglets and sows. *Trudy vsesoyuznogo instituta eksperimental'noi veterinarii*, **23**, 219–239. (Russian.)